T0214016

The IMA Volumes in Mathematics and its Applications
Volume 156

Series Editor:

Fadil Santosa, *University of Minnesota, MN, USA*

For further volumes:
http://www.springer.com/series/811

Institute for Mathematics and its Applications (IMA)

The Institute for Mathematics and its Applications was established by a grant from the National Science Foundation to the University of Minnesota in 1982. The primary mission of the IMA is to foster research of a truly interdisciplinary nature, establishing links between mathematics of the highest caliber and important scientific and technological problems from other disciplines and industries. To this end, the IMA organizes a wide variety of programs, ranging from short intense workshops in areas of exceptional interest and opportunity to extensive thematic programs lasting a year. IMA Volumes are used to communicate results of these programs that we believe are of particular value to the broader scientific community.

The full list of IMA books can be found at the Web site of the Institute for Mathematics and its Applications:

>   http://www.ima.umn.edu/springer/volumes.html.

Presentation materials from the IMA talks are available at

>   http://www.ima.umn.edu/talks/.

Video library is at

>   http://www.ima.umn.edu/videos/.

Fadil Santosa, Director of the IMA

Clint Dawson • Margot Gerritsen
Editors

Computational Challenges in the Geosciences

Springer

Editors
Clint Dawson
Institute for Computational
 Engineering
The University of Texas at Austin
Austin, USA

Margot Gerritsen
Department of Energy Resources
 Engineering
Stanford University
Stanford, CA, USA

ISSN 0940-6573
ISBN 978-1-4939-4576-4   ISBN 978-1-4614-7434-0 (eBook)
DOI 10.1007/978-1-4614-7434-0
Springer New York Heidelberg Dordrecht London

Mathematics Subject Classification (2010): 35R30, 35Q35, 65M06, 65M08, 65M22, 65M32, 65M60, 65M25, 65M99, 74N30, 76S05, 76V05, 80A30, 86A05, 86A22, 86A32

Printed on acid-free paper

Springer is part of Springer Science+Business Media (www.springer.com)

Foreword

This volume is based on the workshop "Societally Relevant Computing" held at the Institute for Mathematics and Its Applications (IMA) at the University of Minnesota, April 11–15, 2010. We would like to thank all the participants for making this a stimulating and productive workshop.

In particular, we would like to thank Michael Brenner, Clint Dawson, Margot Gerritson, and Anna Michalak for their role as workshop organizers and Clint Dawson and Margot Gerriston for organizing this volume, which focuses on computational challenges in the geosciences.

We also take this opportunity to thank the National Science Foundation for its support of the IMA.

Minneapolis, MN

Fadil Santosa
Jiaping Wang

Preface

The Institute for Mathematics and Its Applications (IMA) at the University of Minnesota held a special year on "Simulating Our Complex World: Modeling, Computation and Analysis" in 2010–2011, co-chaired by Susanne Brenner of Louisiana State University and Clint Dawson of the University of Texas at Austin. As part of this special year a workshop was held April 11–15, 2011 on "Societally Relevant Computing," co-organized by Michael P. Brenner of Harvard University; Clint Dawson; Margot Gerritsen of Stanford University; and Anna M. Michalak of the University of Michigan. This volume presents and summarizes some of the cutting-edge research problems specific to geoscience applications that were discussed at this workshop.

Simulation and computation play a critical role in geoscience problems. Examples include the role of anthropogenic emissions on climate and ocean circulation; the prediction of earthquakes and tsunamis; the prediction of paths and storm surges of hurricanes; and designing infrastructure that is capable of withstanding disasters, such as floods and terrorist attacks. These systems exhibit extreme complexity, and there are myriad critical issues that must be accurately addressed. Models need to encompass large numbers of physical effects. All of these problems represent grand challenge computer problems that require pushing the limits of technology, with regard to algorithms and machines and to the development of the physical models themselves.

Critical issues include the development and coupling of algorithms for multiphysics, multiscale applications, verification and validation of these computer models, and quantifying their predictive reliability given potentially large uncertainty in numerical accuracy, code reliability and ultimately the models themselves. How good is good enough? For any complex phenomenon, more and more physical effects can always be included as data and computing power increase, and larger computations can always be designed. However there will always be substantial uncertainty and numerical error to overcome.

While no single volume could address all of these complexities, the contributors to this volume represent a cross-section of geoscience research. Specifically the topics addressed herein include mathematical and numerical complexities of ocean modeling, fast algorithms for Bayesian inversion in geostatistical applications, the modeling of complex multiphase fluids in porous media, real-time storm surge forecasting in the coastal ocean, and modeling and software development for geophysical fluid dynamics applications ranging from tsunamis to debris flows.

We thank the authors for their important contributions to the workshop and to this volume. We thank all who attended the workshop for their participation in the lively discussions that were held. We also thank the IMA for hosting this workshop and for the important work they do in supporting and promoting mathematical research.

Austin, TX   Clint Dawson
Stanford, CA   Margot Gerritsen

Contents

Physical and Computational Issues in the Numerical Modeling of Ocean Circulation

Robert L. Higdon

Abstract The large-scale circulation of the world's oceans can be modeled by systems of partial differential equations of fluid dynamics, as scaled and parameterized for oceanic flows. This paper outlines some physical, mathematical, and computational aspects of such modeling. The topics include multiple length, time, and mixing scales; the choice of vertical coordinate; properties of the shallow water equations for a single-layer fluid, including effects of the rotating reference frame; a statement of the governing equations for a three-dimensional stratified fluid with an arbitrary vertical coordinate; time-stepping and multiple time scales; and various options for spatial discretizations.

Keywords Numerical modeling of ocean circulation • isopycnic and hybrid coordinates • shallow water equations • layered ocean models • barotropic–baroclinic time splitting • spatial grids on a spheroid

Mathematical Subject Classification (2010): 35Q35, 65M99, 76M99, 86A05.

1 Introduction

The circulation of the world's oceans plays a major role in the global climate system, and numerical models of that system typically include models of atmospheric circulation, oceanic circulation, sea ice dynamics, land effects, and biogeochemistry. Local circulation in near-shore regions is also of substantial scientific and societal interest. As a result, numerous numerical models of ocean circulation have been

R.L. Higdon (✉)
Department of Mathematics, Oregon State University, Corvallis, OR 97331-4605, USA
e-mail: higdon@math.oregonstate.edu

C. Dawson and M. Gerritsen (eds.), *Computational Challenges in the Geosciences*, The IMA
Volumes in Mathematics and its Applications 156, DOI 10.1007/978-1-4614-7434-0_1,
© Springer Science+Business Media New York 2013

developed in recent decades. These all involve the numerical solution of systems of partial differential equations of fluid dynamics, as scaled and parameterized for oceanic flows.

The goal of the present paper is to outline some of the physical and computational issues that are encountered when ocean circulation is simulated numerically and to suggest some areas of possible improvement that could engage mathematical scientists. The starting point is to describe some of the space, time, and mixing scales that are found in the ocean, and this is done in Sect. 2. In Sect. 3 we discuss some possible choices for a vertical coordinate in an ocean model; the elevation z is a natural choice, but other possibilities have also been developed. Section 4 describes some aspects of governing equations for oceanic flows, both the shallow water equations for a homogeneous (single-layer) fluid and some equations that model three-dimensional stratified flows. Some aspects of time discretization and space discretization are outlined in Sects. 5 and 6, respectively.

This author has previously written a lengthy survey [6] on the subject of numerical modeling of ocean circulation. The present paper is based on the author's presentation at an IMA workshop on Societally Relevant Computing during April 2011, and it can serve as an update and relatively brief "front end" to the longer, earlier review.

In the sections that follow, some of the background information about ocean physics and ocean modeling practice is stated without explicit citations. The earlier paper [6] includes a lengthy bibliography, and the reader can refer to [6] for further details and references.

2 Multiple Scales

The physical processes in the ocean exhibit a wide range of space, time, and mixing scales. This multiscale character has a profound effect on the fluid flows that are seen and on the numerical algorithms that are used to model them.

2.1 Length Scales

Ocean circulation patterns include wind-driven horizontal gyres; these are closed loops of circulation that span the widths of ocean basins, which are thousands of kilometers wide. An oceanic gyre typically includes an intense boundary current along the western edge of the basin, with a boundary current width on the order of 10^2 km. Examples of western boundary currents include the Gulf Stream in the North Atlantic and the Kuroshio in the North Pacific. In general, ocean currents are not smooth, but instead they meander and shed eddies having widths in the range of tens to hundreds of kilometers. These horizontal length scales are very different from the vertical scales that are present; in the mid ocean the depth is a few kilometers (e.g., four to six), and for surface currents the kinetic energy is mostly confined to the upper few hundred meters.

For large-scale features of ocean circulation, the vertical length scale is thus far less than the horizontal length scale, and this property is known as the *shallow water condition*. It can be shown that the shallow water condition implies that the fluid is approximately hydrostatic, i.e., vertical accelerations are small. (See Sect. 4.1.1.) This can then lead to substantial simplifications in the system of governing equations.

However, not all of the physically relevant motions in the ocean satisfy the shallow water condition. An example arises in the thermohaline circulation, which is also known as the meridional overturning circulation, or global conveyor belt. In this circulation, a portion of the warm water of the Gulf Stream separates from the horizontal gyre and flows into the far northern Atlantic. This water then becomes colder and saltier due to atmospheric forcing, and due to its increased density the water sinks into the deep ocean in narrow convective plumes to form part of a complex three-dimensional global circulation pattern having a period of centuries. The plumes do not satisfy the shallow water condition, and the deep convection is not hydrostatic. This convection is not resolved explicitly in present-day ocean climate models, but instead it is parameterized by some means, such as vertical diffusion. An interesting computational challenge would be to embed nonhydrostatic high-resolution models of the deep-convection zones into a lower-resolution hydrostatic global model. This process could be made difficult by the need to couple the very different dynamics of the two regimes.

2.2 Velocity and Time Scales

A wide range of velocities is represented in the fluid motions and wave motions that are seen in the large-scale circulation of the ocean. For the horizontal component of fluid velocity, a value of $1\,\mathrm{m/s}$ is typical of strong currents, but elsewhere in the ocean the horizontal velocities are considerably smaller. In nearly hydrostatic regions the vertical fluid velocity is typically far smaller than the horizontal velocity, with values such as $10^{-3}\,\mathrm{m/s}$ or less.

Distinct from this are the wave motions that are seen in the large-scale circulation. Here, we are not referring to the familiar surface waves that are readily visible to an observer, but instead we refer to motions having wavelengths that are long relative to the ocean depth. These motions can be classified as *external waves* and *internal waves*, and within each class are *gravity waves* and *Rossby waves*. The review paper [6] includes a derivation and discussion of this modal structure, for the case of linearized flow in a spatial domain having a level bottom.

To visualize these waves, imagine the ocean as a stack of layers separated by surfaces of constant density. If the state of the fluid consists entirely of an external wave, then at any time and horizontal location the horizontal fluid velocity is nearly independent of depth, and all layers are thickened or thinned by approximately the

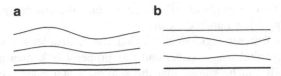

Fig. 1 Illustrations of an external wave (**a**) and an internal wave (**b**). The figures are not to scale, as the depth of the fluid is much smaller than the horizontal length scale, and the vertical displacements are exaggerated in the figures

same proportion. The state of the free surface at the top of the fluid thus displays the mass distribution within the ocean, so the fluid motion at the upper boundary reveals nearly everything about the motion within the fluid. On the other hand, in the case of an internal wave the free surface remains nearly level; the wave motion is characterized by undulations of density surfaces within the fluid, and the mass-weighted vertical average of horizontal velocity is nearly zero. These motions are illustrated in Fig. 1.

In the case of external gravity waves, the wave speed is approximately $\sqrt{gH}$, where H is the depth of the ocean and g is the acceleration due to gravity. For example, if $H = 4{,}000$ m, then the wave speed is approximately 200 m/s. On the other hand, for internal gravity waves the speeds are at most a few meters per second. This large difference in wave speeds can be explained by the fact that for external gravity waves the restoring force is due to the density contrast between water and air, whereas for internal waves the restoring force is due to the far weaker density contrasts within the ocean.

In the case of Rossby waves (Sect. 4.1.4) the restoring mechanism is due to vorticity instead of gravity, and these waves are generated by variations in the Coriolis parameter with latitude and/or variations in the elevation of bottom topography. Rossby waves are typically far slower than gravity waves.

Gravity waves participate in adjustment processes, such as those that occur when a shift in prevailing winds causes a shift in quasi-balanced states in the ocean. Rossby waves are involved in the development of large-scale circulation systems.

To very good approximation, the fast external motions can be modeled by neglecting the density variations within the fluid and then applying the two-dimensional shallow water equations that are discussed in Sect. 4.1. On the other hand, the slow internal motions are fully three-dimensional. These properties can be exploited to develop time-stepping methods that can handle efficiently the multiple time scales that are inherent in large-scale oceanic flows, as discussed in Sect. 5.

2.3 Mixing Scales

A major purpose of an ocean model is to simulate the transport of tracers such as temperature, salinity, and various chemical and biological species. Temperature and salinity directly affect the density of the fluid and thus its dynamics, so temperature and salinity are referred to as active tracers. Other tracers simply move with the flow and are labeled as passive tracers.

Tracers are transported in part by the bulk movement of currents, but they are also mixed by smaller-scale turbulent motions having length scales down to centimeters or less. The full extent of these motions cannot be resolved on the spatial grids of numerical ocean models, so the large-scale effects of these small-scale motions must be parameterized by some sort of diffusion operator.

The nature of this mixing depends on the location within the ocean. In the ocean's interior, the fluid is stratified, and the mixing is highly anisotropic. Empirical observations indicate that at every point in the interior, the plane of strongest mixing is nearly horizontal, and the rate of diffusion in that plane is several orders of magnitude greater than the rate of diffusion perpendicular to that plane. This phenomenon has been explained in terms of buoyancy. The plane of strongest mixing is said to be a neutral plane, and if a fluid parcel were to leave this plane, then buoyancy forces would tend to restore it to the plane; on the other hand, motion within the plane experiences no such impediment and thus is far more active. The plane of strongest mixing need not be exactly horizontal, due to lateral variations of the density of the fluid. The relatively small transport across such a plane is known as *diapycnal* transport.

The preceding argument does not apply to turbulent layers along the boundary of the fluid domain. Of particular interest is the mixed layer at the top of the ocean, for which the mixing is due to wind forcing and convective overturning. At any location the structure of the mixed layer may vary with time, due to weather and seasonal variations and to heating and cooling throughout the diurnal cycle. The mixed layer is generally tens of meters thick, although it can be thousands of meters thick in locations of deep convection. The mixed layer provides the means of communication between ocean and atmosphere, so its representation is an important part of an ocean model.

3 The Choice of Vertical Coordinate

For the vertical coordinate in a three-dimensional ocean circulation model, a traditional choice has been the elevation z. This coordinate is used, for example, in the widely used Bryan–Cox class of ocean models (e.g., Griffies [4]). Another possibility is a terrain-fitted vertical coordinate σ, which measures the relative position between the top and bottom of the fluid domain. The top and bottom correspond to constant values of σ, so this framework is natural for representing irregular bottom topography; see Fig. 2. The σ-coordinate system is widely used in modeling the ocean circulation in near-shore regions (e.g., Shchepetkin and McWilliams [10]).

Fig. 2 A terrain-fitted vertical coordinate. The shaded region represents variable bottom topography, and the curves are cross-sections of surfaces of constant vertical coordinate σ

3.1 Isopycnic Coordinates

A third possibility is to use an *isopycnic coordinate*, which is a quantity related to density. A vertical discretization would then divide the fluid into layers having distinct physical properties, as opposed to using a geometric discretization such as one found with z or σ.

A specific possibility for an isopycnic coordinate is *potential density*. To define that quantity, imagine that a fluid parcel is moved to a reference pressure, without any heat flow into or out of the parcel and without any changes in the composition of the fluid in that parcel. The resulting density is then the potential density, relative to that reference pressure. The reference pressure could be atmospheric pressure, or it could be the pressure at some specified depth. However, if atmospheric pressure is used as the reference pressure, then the resulting potential density could be nonmonotonic in z in regions of weak stratification, which is not suitable for a vertical coordinate. A reference pressure at a specified depth is therefore used instead. Section 4.2.2 includes an argument for using potential density instead of true density as a vertical coordinate.

The use of an isopycnic coordinate is motivated by the properties of oceanic mixing that are discussed in Sect. 2.3. For modeling the ocean's interior, an ideal coordinate would be a vertical coordinate s whose level surfaces are neutral surfaces. That is, at each point on a surface of constant s, the tangent plane is the plane of strongest mixing, which was discussed previously; see Fig. 3. (Such a plane is commonly described as the tangent plane to the surface of constant potential density, where the potential density is referenced to the pressure at that particular point.) Suppose that a neutral coordinate s is used, and assume that the fluid domain is discretized into layers that are separated by surfaces of constant s. The strong lateral mixing then occurs only within coordinate layers, and the only transport of tracers between layers is the much weaker diapycnal diffusion that is described in Sect. 2.3. The relatively warm upper layers are then distinguished automatically from the colder lower layers by the choice of coordinate system. This can be an advantage for long-term simulations, such as those that arise in climate modeling.

More specifically, the transport of tracers is modeled by advection–diffusion equations, and with such equations the approximations to the horizontal advection terms typically generate spurious numerical diffusion. This diffusion is directed along surfaces of constant vertical coordinate. In the scenario described here, this diffusion is confined to the coordinate layers and does not mask the subtle diapycnal

Fig. 3 Isopycnic coordinate. The solid curve represents a cross-section of a surface of constant isopycnic coordinate. Ideally, a tangent plane to such a surface would be the plane of strongest mixing at that point. Spurious numerical diffusion in horizontal advection terms would then be confined to coordinate layers and would not mask the subtle physical diffusion between layers

diffusion that occurs between layers. Representing the latter is then under the direct control of the modeler. On the other hand, if the vertical coordinate is z or σ, then the coordinate surfaces can cross the boundaries of the physical layers, and the numerical diffusion can generate nonphysical transport of tracers between those layers.

Unfortunately, the idea of isopycnic coordinate has limitations. (1) In general, exact neutral surfaces do not exist; the local picture of neutral planes does not extend globally. Instead, one must choose an approximation to this idea, and potential density (referenced to a suitable fixed pressure) is believed to be a workable proxy. (2) Isopycnic coordinate surfaces can intersect the bottom of the fluid domain, and they can also intersect the top of the fluid, due to lateral variations in temperature. This implies that the thicknesses of certain layers can tend to zero, and thus some vertical resolution is lost in such locations. (3) The physical conditions that motivate the use of an isopycnic coordinate are valid in the oceanic interior, but not in turbulent boundary layers. Using an isopycnic coordinate over the entire ocean is thus not fully justified, on physical grounds.

3.2 Hybrid Coordinates

An alternative to using purely z, σ, or an isopycnic coordinate is to use a hybrid of these three. In the case of a hybrid coordinate, an isopycnic coordinate is used in the ocean's interior, where its use is justified by the reasoning given above. However, in vertically mixed regions near the upper boundary, the coordinate may depart from isopycnic and become z instead. For near-shore regions, an option that is sometimes used is to transform the isopycnic surfaces into σ-coordinate surfaces. A hybrid-coordinate strategy for the ocean is described by Bleck [1], and some recent improvements are described by Bleck et al. [2] in the analogous context of atmospheric modeling.

An example of a hybrid coordinate system is shown in Fig. 4. The figure shows a vertical cross-section of the southern Atlantic Ocean, as computed with the Hybrid Coordinate Ocean Model [1]. This cross-section shows the top 1,000 m of the fluid domain, and it is taken at fixed longitude 54.9° W, with the latitude varying from approximately 36° S to 56.5° S. The thin, numbered curves are cross-sections of surfaces of constant potential density, as referenced to the pressure at a depth of 2,000 m (or, more precisely, to 2,000 decibars). The values of potential density are

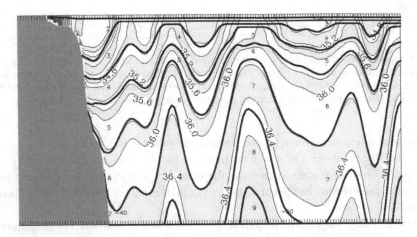

Fig. 4 Example of a hybrid vertical coordinate. The figure shows a vertical cross-section of the southern Atlantic Ocean, for fixed longitude, as obtained from a numerical simulation. The thin curves are cross-sections of surfaces of constant potential density, and the thick curves are cross-sections of surfaces of constant vertical coordinate. In the ocean's interior, the vertical coordinate is isopycnic, and near the upper boundary the coordinate is z. In this figure the vertical extent is 1 km, and the horizontal extent is approximately 2,300 km. (Figure provided by Rainer Bleck, NASA Goddard Institute for Space Studies)

obtained by adding 1,000; for example, the curve labeled 36.0 represents potential density $1036.0 \, \text{kg}/\text{m}^3$. The thicker curves are cross-sections of surfaces of constant vertical coordinate. In the ocean's interior these align with the potential density surfaces, so the coordinate is isopycnic in that region. However, near the upper boundary of the ocean, the coordinate surfaces are level.

The use of hybrid coordinates is relatively recent, compared to the other coordinate systems described here, and it presents some physical and algorithmic challenges. In regions of the upper ocean where vertical mixing causes a loss of stratification, isopycnic layers lose their identity. In such regions the vertical coordinate cannot be isopycnic, so an alternative should be used. In addition, a given range of the isopycnic variable (e.g., potential density) may exist in some regions of the ocean but not in others; in locations where an isopycnic layer does not exist physically, its mathematical representation deflates to zero thickness. It is therefore necessary to employ a grid generator that continually manages the structure of the vertical grid, which can evolve with time. For the sake of providing vertical resolution, the grid generator should maintain a minimum layer thickness for all of the coordinate layers, except perhaps those that intersect the bottom topography. On the other hand, each coordinate surface should represent its "target" value of the isopycnic variable whenever physical conditions permit, i.e., the fluid is stratified at that location, and the target value is within the range that is physically present.

Depending on its design, it is possible that a grid generator could produce nonphysical motions of coordinate surfaces relative to the fluid. Such motions would be represented in the vertical transport terms in the advection–diffusion equations for the transport of tracers (see Sects. 4.2.1 and 4.2.2), and the spurious diffusion inherent in numerical advection schemes could then compromise the water properties in the affected regions. The vertical coordinate system should therefore be managed in a way that minimizes such coordinate movements, while maintaining adequate grid resolution in vertically mixed regions and maximizing the volume of the ocean that is represented with isopycnic coordinates.

4 Governing Equations

The survey [6] contains a detailed derivation of the equations for conservation of mass, momentum, and tracers for a hydrostatic fluid that is in motion relative to a rotating spheroid. In that derivation the vertical coordinate is arbitrary, and the horizontal coordinates are arbitrary orthogonal curvilinear coordinates.

The goal of the present section is to describe some aspects of these equations, without going into full details or derivation; the reader is referred to [6] for further information. Section 4.1 describes the two-dimensional shallow water equations for a homogeneous (single-layer) fluid, and Sect. 4.2 outlines the three-dimensional equations. For notational simplicity, the horizontal coordinates are taken to be Cartesian coordinates, so the equations describe motion relative to a tangent plane attached to the rotating spheroid. (At every point on the spheroid, the terms "horizontal" and "vertical" refer to directions that are tangent and normal to the spheroid, respectively.)

4.1 Shallow Water Equations

The shallow water equations are based on the assumption that the fluid has constant density and that the fluid satisfies the shallow water condition, i.e., the depth of the fluid is far less than the horizontal length scale for the motions of interest.

The assumption of constant density is used in certain applications, such as to model tides, tsunamis, and storm surge. However, this assumption is too restrictive for modeling the general circulation of the ocean. Nonetheless, in studies of the general circulation, the shallow water equations are still of great interest, for multiple reasons. (1) These equations enable one to demonstrate, in a relatively simple setting, some of the effects of a rotating reference frame, such as geostrophic balance, conservation of potential vorticity, and Rossby waves. (2) For purposes of numerical simulations, it is highly desirable to split the full three-dimensional problem into fast and slow subsystems that are solved by different techniques. In such a splitting, the fast subsystem is very similar to the shallow water equations. (3) For a three-dimensional system with an isopycnic vertical coordinate, a vertical discretization resembles a stack of shallow-water models.

4.1.1 Implications of the Shallow Water Condition

It was mentioned in Sect. 2.1 that the shallow water condition implies that the fluid is nearly hydrostatic. This conclusion can be developed via formal scaling arguments, but here we only summarize the main ideas with an informal discussion. If the horizontal velocity field displays positive divergence, then the fluid is being spread out in the horizontal, which implies that the free surface at the top of the fluid must fall. Similarly, negative horizontal divergence implies that the free surface must rise. Now assume that the shallow water condition is valid, and consider a horizontal distance L over which the horizontal velocity varies significantly. The horizontal divergence acts on the mass flux over the relatively small vertical extent of the fluid, but the effects of this divergence are spread out over the much greater distance L. Thus the free surface elevation must change very slowly. The vertical acceleration in the fluid is then very small, i.e., the fluid is nearly in hydrostatic balance.

The case of exact balance can be formulated as follows. Let $p(x,y,z,t)$ denote the pressure in the fluid at position (x,y,z) and time t, with z increasing upward; let ρ denote the density of the fluid; and let g denote the acceleration due to gravity. (In the present paragraph, it is immaterial whether ρ is constant.) Suppose that for each subregion V of the fluid, the vertical component of the net pressure force on the fluid in V is balanced exactly by the weight of that volume of fluid. A consideration of surface integrals and the divergence theorem then implies that $\partial p/\partial z = -\rho g$ everywhere in the fluid.

For the sake of the following, let $\eta(x,y,t)$ denote the elevation of the free surface, relative to the equilibrium state $\eta = 0$. Also let p_0 denote the atmospheric pressure at the top of the fluid, and assume p_0 is constant.

Remark 4.1. The shallow water condition and the assumption of constant density reduce the system to two horizontal dimensions (plus time), provided that the horizontal velocity is independent of z at some time.

The reason for this is as follows. The hydrostatic condition $\partial p/\partial z = -\rho g$, together with the assumption of constant ρ, implies that the pressure in the interior of the fluid is $p(x,y,z,t) = p_0 + \rho g(\eta(x,y,t) - z)$. Thus, for a fixed elevation z within the fluid, lateral variations in pressure are due entirely to variations in the elevation of the free surface at the top of the fluid. Furthermore, the nature of that variation is independent of z. Horizontal fluid acceleration is thus independent of z; if the horizontal velocity is independent of z at some time t_0, it must therefore be independent of z for all times. The components of horizontal velocity are thus functions of (x,y,t), and the flow can described completely by this velocity and the elevation η.

4.1.2 Statement of the Shallow Water Equations

In keeping with the preceding conclusions, let $\mathbf{u}(x,y,t) = (u(x,y,t), v(x,y,t))$ denote the horizontal velocity of the fluid; more precisely, $\mathbf{u}(x,y,t)$ is the velocity at time

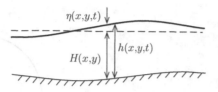

Fig. 5 Illustration of the mass variables in the shallow water equations. The upper solid curve is a cross-section of the free surface at the top of the fluid, the lower solid curve represents bottom topography, and the horizontal dashed line represents the free surface when the fluid is at rest

t of whatever fluid parcel happens to be at position (x,y) at that time. Furthermore, $\mathbf{u}(x,y,t)$ is the velocity relative to a *rotating* coordinate system, the equations that follow describe motion relative to a tangent plane on a rotating spheroid; $\mathbf{u}(x,y,t)$ is not a velocity relative to an inertial reference frame. Also, let $h(x,y,t)$ denote the thickness of the fluid layer, i.e., the vertical distance between the top and bottom of the fluid, and let $H(x,y)$ denote the thickness of the layer when the fluid is at rest. Then $\eta(x,y,t) = h(x,y,t) - H(x,y)$. These variables are illustrated in Fig. 5. Due to the possibility of variable bottom topography, H need not be constant.

The equation for conservation of mass is

$$\frac{\partial h}{\partial t} + \frac{\partial}{\partial x}(uh) + \frac{\partial}{\partial y}(vh) = 0, \tag{4.1}$$

or $\partial h/\partial t + \nabla \cdot (\mathbf{u}h) = 0$. Here, the layer thickness h plays the role of the mass variable; the mass of the water column on any horizontal region A is $\int_A \rho h(x,y,t)\,dx\,dy$, so ρh is the mass per unit horizontal area in the fluid layer.

The equations for conservation of momentum can be written in component form as

$$\frac{Du}{Dt} - fv = -g\frac{\partial \eta}{\partial x}, \tag{4.2}$$

$$\frac{Dv}{Dt} + fu = -g\frac{\partial \eta}{\partial y}, \tag{4.3}$$

or in vector form as $D\mathbf{u}/Dt + f\mathbf{u}^{\perp} = -g\nabla\eta$. Here, f is the Coriolis parameter (4.4); $\mathbf{u}^{\perp} = (-v,u)$; and $D\mathbf{u}/Dt = \mathbf{u}_t + u\mathbf{u}_x + v\mathbf{u}_y$, where subscripts denote partial differentiation.

The operator D/Dt is the material derivative, i.e., the time derivative as seen by an observer moving with the fluid. To see this, let $(x(t),y(t))$ denote the position of such an observer. The velocity of the observer is then $(x'(t),y'(t)) = (u(x(t),y(t),t),v(x(t),y(t),t))$, or $\mathbf{u}(x(t),y(t),t)$. Now let $q(x,y,t)$ be a quantity defined on the fluid region, such as a component of velocity or the density of a tracer. The value of q that is seen by the observer is $q(x(t),y(t),t)$, and the time derivative of this quantity is $q_t + uq_x + vq_y = Dq/Dt$.

Fig. 6 The Coriolis
parameter $f = 2\mathbf{\Omega} \cdot \mathbf{k}$ is the
local vertical component of
the planetary vorticity $2\mathbf{\Omega}$

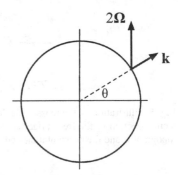

The material derivative $D\mathbf{u}/Dt$ is the time derivative of velocity as seen by an observer moving with the fluid, and so it is the acceleration of that observer and thus the acceleration of a fluid parcel. However, this is the acceleration relative to the *rotating* reference frame of the spheroid. The Coriolis term $f\mathbf{u}^{\perp} = (-fv, fu)$ provides the necessary correction so that the sum $D\mathbf{u}/Dt + f\mathbf{u}^{\perp}$ gives the acceleration relative to an inertial reference frame, and it is in an inertial frame where one can apply Newton's second law to obtain a statement of conservation of momentum. The term $-g\nabla\eta$ provides the pressure forcing that produces the acceleration, and it reflects the fact that the lateral variations of pressure within the fluid are due to lateral variations in the elevation of the free surface.

4.1.3 Coriolis Term and Geostrophic Balance

The value of the Coriolis parameter f is defined as follows. Let Ω denote the rate of rotation of the spheroid, in terms of radians per unit time. If a fluid is stationary relative to this spheroid, then the vorticity (curl of the velocity field) is the *planetary vorticity* $2\mathbf{\Omega}$. Here, the velocity is measured relative to an inertial frame attached to the axis of the spheroid, and $\mathbf{\Omega}$ is a vector with magnitude $\Omega = |\mathbf{\Omega}|$ that is aligned with the axis so that $\mathbf{\Omega}$ and the direction of rotation satisfy the right-hand rule. At every point on the rotating spheroid, the Coriolis parameter is the local vertical component of the planetary vorticity $2\mathbf{\Omega}$. That is,

$$f = 2\mathbf{\Omega} \cdot \mathbf{k} = 2\Omega \sin\theta, \tag{4.4}$$

where $\mathbf{k}$ is the unit upward vector at the point in question, and θ is the latitude. See Fig. 6. At the equator, $f = 0$, so rotation does not play a role in the momentum equations (4.2) and (4.3) in that case. At the North and South Poles, f equals 2Ω and -2Ω, respectively.

For large-scale motions of the ocean (and atmosphere), there is usually an approximate balance between the Coriolis terms and the lateral pressure forcing, except near the equator. This balance is called the *geostrophic balance*. In the case of an exact balance, the momentum equations (4.2) and (4.3) reduce to $f\mathbf{u}^{\perp} = -g\nabla\eta$,

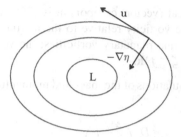

Fig. 7 Geostrophic balance in the shallow water equations. The velocity is orthogonal to the pressure forcing, and curves of constant free-surface elevation are streamlines for the fluid flow. This behavior is due to the strong influence of the rotating reference frame. In this figure, the center is assumed to be a local minimum of the elevation, and $f > 0$

or

$$-fv = -g\frac{\partial \eta}{\partial x},$$

$$fu = -g\frac{\partial \eta}{\partial y}.$$

The left side of this system is orthogonal to the velocity vector $\mathbf{u} = (u,v)$, so $u\eta_x + v\eta_y = 0$, or $\mathbf{u} \cdot \nabla\eta = 0$. That is, the velocity is orthogonal to the pressure forcing. Equivalently, the velocity is tangent to curves of constant η, so that the level curves for the free-surface elevation are streamlines for the fluid flow. See Fig. 7.

This result is counterintuitive, based on our everyday experience, but it is a feature of flows that are dominated by rotation. For an analogy, suppose that you want to walk a straight path on a rotating merry-go-round. This path is curved relative to the ground, so you need to exert a side force in order to remain on that straight/curved path.

4.1.4 Potential Vorticity and Rossby Waves

The horizontal velocity $\mathbf{u} = (u,v)$ is the velocity relative to the rotating reference frame, and the vorticity relative to this rotating frame is then

$$\operatorname{curl} \mathbf{u} = \nabla \times \mathbf{u} = \left(\frac{\partial v}{\partial x} - \frac{\partial u}{\partial y}\right)\mathbf{k} = (v_x - u_y)\mathbf{k},$$

where again $\mathbf{k}$ is the unit upward vector. The quantity $\zeta = v_x - u_y$ is twice the rate of fluid rotation about the axis $\mathbf{k}$, as viewed in the rotating reference frame. As mentioned in Sect. 4.1.3, the planetary vorticity 2Ω is the curl of the velocity field associated with a rigid rotation of the spheroid about its axis, viewed in the inertial frame, and its local vertical (vector) component is $(2\mathbf{\Omega} \cdot \mathbf{k})\mathbf{k} = f\mathbf{k}$. The *absolute vorticity* of the fluid is the vorticity relative to the inertial reference frame, and it is the sum of the relative and planetary vorticities; the vertical component of the absolute vorticity is then $(\zeta + f)\mathbf{k}$.

Remark 4.2. Some manipulations of the mass and momentum equations (4.1)–(4.3) yield

$$\frac{D}{Dt}\left(\frac{\zeta + f}{h}\right) = 0. \tag{4.5}$$

The quantity $q = (\zeta + f)/h$ is known as the *potential vorticity*. Since the horizontal velocity of the fluid is independent of z, the fluid moves in columns, and Eq. (4.5) states that the potential vorticity for each material column remains constant as the column moves with the flow.

For example, suppose that h remains constant while a water column changes latitude. Northward movement implies that f increases, so ζ must decrease; this then imparts a clockwise rotation in the fluid, relative to the rotating reference frame. Similarly, southward movement imparts a counterclockwise rotation. These relative rotations provide restoring mechanisms for the lateral movement of water columns, and the resulting waves are known as Rossby waves. Similarly, variations in bottom topography can cause changes in the layer thickness h, and this then induces changes in $\zeta + f$ via vortex stretching.

4.2   Governing Equations in Three Dimensions

Now consider a hydrostatic fluid for which the density is *not* constant, and assume that for each (x, y, t) the density ρ decreases monotonically as z increases. Because the fluid is hydrostatic, the pressure p must also decrease monotonically with z. Within the fluid, let s be a generalized vertical coordinate; for example, s could be z, σ, an isopycnic coordinate, or a hybrid of these three. Assume that s is strictly monotone increasing as z increases, for each (x, y, t); in order to satisfy this condition in the isopycnic case, s could be the reciprocal of potential density.

All dependent variables are then functions of (x, y, s, t). The surfaces of constant s need not be horizontal; however, the horizontal coordinates x and y shall measure distances projected onto a horizontal plane, not along surfaces of constant s. Similarly, the velocity components $u(x, y, s, t)$ and $v(x, y, s, t)$ measure velocity that is truly horizontal.

4.2.1   Conservation of Mass

In the present general setting, it is not appropriate to use the density ρ as the dependent variable for mass, since in the isopycnic case, ρ is essentially an independent variable. Instead, for a mass variable use

$$-\frac{\partial p}{\partial s} = -p_s = |p_s|.$$

(As s increases, p must decrease, so $\partial p/\partial s \leq 0$.) For an interpretation of this quantity, consider two coordinate surfaces defined by $s = s_0$ and $s = s_1$, where $s_0 > s_1$. The size of $-\partial p/\partial s$ indicates the amount of mass in the layer bounded by those surfaces. In particular,

$$\int_{s_1}^{s_0} -p_s(x,y,s,t)\,\mathrm{d}s = p(x,y,s_1,t) - p(x,y,s_0,t) = \Delta p(x,y,t) \geq 0,$$

where Δp denotes the vertical pressure difference between the two coordinate surfaces. Since the fluid is assumed to be hydrostatic, Δp is the weight per unit horizontal area in the layer, i.e., g times the mass per unit horizontal area.

The equation for conservation of mass can be written in the form

$$\frac{\partial}{\partial t}(p_s) + \frac{\partial}{\partial x}(up_s) + \frac{\partial}{\partial y}(vp_s) + \frac{\partial}{\partial s}(\dot{s}p_s) = 0. \tag{4.6}$$

Here, $\dot{s} = Ds/Dt$, the time derivative of s following fluid parcels. The last term on the left side of Eq. (4.6) is an analogue of the preceding two, since $u = \dot{x}$ and $v = \dot{y}$.

For an example of Eq. (4.6), let s be the elevation z. Then $\dot{s} = \dot{z} = Dz/Dt$, which is the vertical component w of linear velocity. Also, $p_s = p_z = -\rho g$, by the hydrostatic condition. The mass equation (4.6) then becomes

$$\frac{\partial \rho}{\partial t} + \frac{\partial}{\partial x}(u\rho) + \frac{\partial}{\partial y}(v\rho) + \frac{\partial}{\partial z}(w\rho) = 0,$$

as expected.

Equations for the transport of tracers have a form similar to Eq. (4.6), with tracer density replacing p_s, and with the addition of terms that represent the anisotropic diffusion described in Sects. 2.3 and 3.1.

4.2.2 Vertical Discretization

Numerical solution of the governing equations requires that the spatial domain be discretized by some means. For a vertical discretization, partition the fluid into regions separated by the coordinate surfaces $s = s_0, s_1, \ldots, s_R$, with $s_0 > s_1 > \cdots > s_R$. Now integrate the mass equation (4.6) vertically over layer r, i.e., for $s_r < s < s_{r-1}$, to obtain

$$\frac{\partial}{\partial t}(\Delta p_r) + \frac{\partial}{\partial x}(u_r \Delta p_r) + \frac{\partial}{\partial y}(v_r \Delta p_r) + (\dot{s}p_s)_{s=s_r} - (\dot{s}p_s)_{s=s_{r-1}} = 0. \tag{4.7}$$

Here, $\Delta p_r(x,y,t) = p(x,y,s_r,t) - p(x,y,s_{r-1},t)$, and

$$u_r(x,y,t) = \frac{1}{\Delta p_r} \int_{s_r}^{s_{r-1}} u(x,y,s,t)\left(-p_s(x,y,s,t)\right) ds.$$

The quantity u_r is the mass-weighted vertical average of u in layer r. The mass-weighted vertical average v_r is defined similarly. In the preceding, the subscript in the quantity $\dot{s}p_s$ refers to a partial derivative, but otherwise the subscripts refer to layers and interfaces.

The terms involving $\dot{s}p_s$ represent mass transport between layers. In the case $s = z$, this term is simply $\dot{s}p_s = \dot{z}p_z = -w\rho g$. In general, a nonzero value of $\dot{s}$ means that the value of s associated with a fluid parcel is changing with respect to time. In the case of an isopycnic coordinate, such a change could arise from a heating or cooling of the fluid. A fluid parcel could be motionless in space, but due to thermodynamic changes a surface of constant s could move across the parcel; from the point of view of the coordinate surface, the fluid parcel moves across the surface. In the case of a hybrid coordinate, a grid generator might deliberately move a coordinate surface relative to the fluid, and this action would be represented by a nonzero value of $\dot{s}$.

Related ideas explain why one would use potential density instead of true density for an isopycnic coordinate. Suppose that s were the reciprocal of true density ρ. Also suppose that at some horizontal position (x,y), the layer between $s_r = 1/\rho_r$ and $s_{r-1} = 1/\rho_{r-1}$ is thickened due to lateral mass transport and that the effect of this thickening is to move the lower boundary of the layer deeper into the fluid. Due to compressibility, the density of the fluid on this lower boundary must increase, so that its density is no longer equal to ρ_r and its vertical coordinate is no longer equal to $s_r = 1/\rho_r$. Instead, the coordinate surface $s = s_r$ must be located somewhere higher in the fluid; a portion of the water mass must therefore have crossed the surface $s = s_r$. This transport would be included in the term $(\dot{s}p_s)_{s=s_r}$. On the other hand, *potential* density is invulnerable to such effects, since by definition it is linked to a fixed reference pressure; in that case, the transport term $\dot{s}p_s$ can be reserved for subtler thermodynamic effects.

4.2.3   Conservation of Momentum

In the vertically discrete case, the conservation of horizontal momentum in layer r can be expressed by the equation

$$\frac{D\mathbf{u_r}}{Dt} + f\mathbf{u_r}^\perp = -(\nabla M - p\nabla\alpha) + \frac{1}{\Delta p_r}\left[\dot{s}p_s\mathbf{u}\right]_{s=s_r}^{s=s_{r-1}} + \mathbf{F_r} . \tag{4.8}$$

Here, $\mathbf{u_r} = (u_r, v_r)$, $\alpha = 1/\rho$, $M = \alpha p + gz$ is the Montgomery potential, and $\mathbf{F_r}$ denotes the effects of viscosity and shear stresses. The quantity $\dot{s}p_s\mathbf{u}$ represents momentum transport across coordinate surfaces.

The quantity $-(\nabla M - p\nabla\alpha)$ represents the horizontal pressure forcing. The differentiations in the operator $\nabla = (\partial/\partial x, \partial/\partial y)$ are taken for fixed vertical coordinate s. However, surfaces of constant s need not be horizontal, whereas the pressure forcing requires the gradient of pressure for fixed z; the difference of the two terms contains the required transformations. The hydrostatic condition $\partial p/\partial z = -\rho g$ is equivalent to $\alpha \partial p/\partial s + g \partial z/\partial s = 0$, which in turn is equivalent to

$$\frac{\partial M}{\partial s} = p\frac{\partial \alpha}{\partial s}. \tag{4.9}$$

Equation (4.9) can be discretized to communicate pressure effects between layers.

In the case of a layer of constant density, the hydrostatic condition implies that M is independent of vertical position. Suppose that a homogeneous (single-layer) fluid is bounded above by constant atmospheric pressure p_0, and let z be the vertical coordinate. In this case, $M(x,y,t)$ equals its value at the free surface, so $M(x,y,t) = \alpha p_0 + g\eta(x,y,t)$, and therefore $-(\nabla M - p\nabla\alpha) = -g\nabla\eta$. The pressure forcing used in Eq. (4.8) thus reduces, in this case, to the pressure forcing that is used in the shallow water equations.

4.2.4 Summary of the Three-Dimensional Equations

The governing equations for the flow consist of the mass equation (4.7); the momentum equation (4.8); advection–diffusion equations for temperature, salinity, and perhaps other tracers; a discretization of Eq. (4.9); and an equation of state that relates potential density, potential temperature, and salinity. (Potential temperature has a definition analogous to that of potential density.)

In the case of an isopycnic coordinate, the quantity $\dot{s}$ is generally small, and Eqs. (4.7) and (4.8) for conservation of mass and momentum in layer r resemble the shallow water equations. The system of equations for the entire three-dimensional fluid can then be regarded informally as a stack of shallow-water models, with mechanisms for transporting mass and pressure effects between layers. Isopycnic and hybrid-coordinate models are therefore often referred to as *layered models*.

5 Multiple Time Scales and Time-Stepping

As noted in Sect. 2.2, the ocean admits motions that vary on a wide range of time scales. In particular, in the mid ocean the external gravity waves have speeds that are two orders of magnitude greater than the other motions that are seen there. If an explicit time-stepping method is used to solve a system of partial differential equations, then the maximum allowable time step is limited by the fastest motions that are present, and in the present application the fast external gravity waves impose a severe constraint.

As described in Sect. 2.2, the external motions are nearly two-dimensional, in the sense that if the motion of the ocean is purely an external mode, then the motion is described approximately by the behavior of the free surface and the horizontal velocity at the top of the fluid. The irony of this situation is that the circulation of the ocean is complex and three-dimensional, yet the time step for explicit schemes is severely limited by relatively simple two-dimensional dynamics.

5.1 Barotropic–Baroclinic Splitting

A widely used remedy for this difficulty is to split the problem into two subsystems. One subsystem models (approximately) the slow internal and advective motions, and it is solved explicitly with a time step that is determined by these slow motions. This system is referred to as the *baroclinic* system. The other subsystem, known as the *barotropic* system, is used to model (approximately) the fast external motions. This system is two-dimensional and resembles the shallow water equations for a fluid of constant density. The barotropic equations can be solved either explicitly with a short time step dictated by the fast external gravity waves, or implicitly with the same long time step that is used for the baroclinic equations. Using small time steps or implicit methods raises the specter of high computational cost, but these measures are applied to a relatively simple two-dimensional subsystem instead of the full three-dimensional system. If an explicit method with short steps is used for the barotropic equations, the resulting coupled algorithm is sometimes referred to as *split-explicit* time-stepping.

An explanation of terminology is in order. In classical fluid dynamics, the term "barotropic" refers to a fluid state in which the pressure is constant along surfaces of constant density. This is approximately the case for a purely external motion, as the undulations of each density surface are synchronized with those of the free surface at the top of the fluid. (See Fig. 1.) However, this is not the case with internal waves, and the term "baroclinic" is used in that situation.

In general, it is not possible to split the fast and slow motions exactly into separate subsystems; instead, approximate splittings must be sufficiently accurate. Different approaches have been used to implement the idea of barotropic–baroclinic splitting, as outlined in [6]. Generally speaking, a splitting proceeds along the following lines.

For the barotropic equations, use vertical summation and/or averaging of the mass and momentum equations. This produces two-dimensional equations in which the vertical variations in the dependent variables have been homogenized. For the three-dimensional baroclinic equations, two basic approaches can be used. For the mass equations, at each time step one could solve the mass equation in each layer and then employ some small flux adjustments to ensure that the algorithms used to solve the layer mass equations are consistent with those used to solve the vertically integrated barotropic mass equation [5]. This has the effect of applying a time filter to the layer equations so that a long time step can be used safely. For the momentum equations, one could use an analogous approach, or an alternative is to subtract the equation for the vertically averaged barotropic velocity from the equation for the velocity in layer r to obtain an equation for the (slowly varying) baroclinic part of the velocity in layer r.

5.2 Remarks on Constructing a Time-Stepping Method

A time integration method that has traditionally been used in ocean modeling is the centered, three-time-level, leap-frog method. However, this method allows a computational mode consisting of sawtooth oscillations in time, and in a nonlinear model this mode can grow and swamp the solution unless it is filtered sufficiently. Accordingly, some recent work has gone into developing alternative time-stepping methods that do not suffer from this limitation. See, e.g., [4, 10]. For example, this author [5] has developed a two-time-level method for use with barotropic–baroclinic splitting for layered models. This method involves some predicting and correcting, is stable subject to the usual Courant–Friedrichs–Lewy condition, displays essentially no spurious numerical dissipation, and is second-order accurate.

For the sake of potential future development in this area, following are some thoughts related to the development of time-stepping methods for ocean circulation models.

1. For reasons of computational efficiency, a time integration method should use a barotropic–baroclinic splitting, in some way or another.
2. In such a splitting, each subsystem requires data from the other. However, somebody has to go first, and this is the reason for the predicting and correcting in the time-stepping method in [5]. The ordering and nature of operations must be chosen so as to maintain numerical stability.
3. Suppose that the barotropic equations are solved explicitly with many short substeps of a (long) baroclinic time interval. When values of the rapidly varying barotropic variables are communicated to provide forcing to the baroclinic equations, time averages should be used instead of instantaneous values in order to avoid problems with sampling and aliasing.
4. Enforcing consistency between the barotropic and baroclinic equations may require some care.
5. The Coriolis terms are energetically neutral, so their implementation should be stable and also avoid numerical dissipation. The Coriolis terms and pressure terms should be evaluated at the same time level; otherwise, the geostrophic balance will contain a first-order truncation error in time.
6. The baroclinic equations are applied in each coordinate layer, and the forcing terms are complicated. The momentum equations include Coriolis terms, momentum flux, pressure forcing, viscosity, and shear stress. The mass equations include lateral mass fluxes and fluxes across coordinate surfaces, and it is likely that flux limiting will be necessary.

7. Algorithms should be feasible for use with many layers, such as dozens of layers or more. For example, the complexity of the forcing may place limitations on the number of stages that can be used in a time-stepping method. Also, the layers are coupled, and it may not be practical to use a Riemann solver over an entire water column to compute mass and momentum fluxes.

8. For the sake of integrations over long time intervals, one could consider implicit time-stepping methods for the full three-dimensional model. Such methods require the solution of large systems of algebraic equations at each time step, and this task is substantial. One possibility for solving such systems is to use a Jacobian-free Newton–Krylov method (e.g., Knoll and Keyes [7]), with a preconditioner based on split-explicit time-stepping.

6 Spatial Discretization

In its horizontal extent, the ocean occupies a complicated domain on the surface of a spheroid. Due to the complexity of oceanic flows, one would like to use as high a spatial resolution as possible; however, for reasons given in Sect. 2.3, there will inevitably be unresolved processes that must be parameterized in some manner. The present section describes some horizontal spatial discretizations that have been used in ocean models and/or may be used in the future.

6.1 Rectangular Grids

Logically rectangular grids have traditionally been used in the numerical modeling of ocean circulation, and in operational ocean modeling they remain the dominant type of grid. Such grids are relatively simple and are natural for using finite difference and finite volume methods.

However, a problem with rectangular grids is the inevitable convergence of grid lines. In the case of a latitude-longitude grid, the grid lines converge at the North and South Poles, and the North Pole is within the fluid domain for a global ocean model. The resulting singularity in the curvilinear coordinates makes this grid unsuitable for such a model.

An alternative that has been used in operational models is a tripolar grid. With such a grid, latitude and longitude are used to the south of a specified latitude. To the north of that latitude, each curve of constant longitude extends northward to connect smoothly with another longitude line that comes northward from another part of the Earth. Orthogonal curves to these quasi-longitude curves then yield a rectangular grid. This arrangement generates two coordinate poles for the northern hemisphere, but the grid can be configured so that these two poles lie on land masses. See Fig. 8.

On a rectangular grid, there are multiple possibilities for the spatial arrangement of the dependent variables, as reviewed, for example, in [4, 6]. In the following

Fig. 8 A tripolar rectangular grid. In the northern hemisphere, the grid has two singular points, and these are located outside the fluid domain. (Figure provided by Philip W. Jones, Los Alamos National Laboratory)

descriptions, regard grid rectangles as mass cells. In the 1960s, Arakawa and coworkers analyzed some possible grid arrangements and labeled these grids A through E. For example, on the A-grid all quantities are defined at cell centers; with the B-grid, the horizontal components of fluid velocity are defined at cell corners; and in the case of the C-grid, the normal components of fluid velocity are defined at the centers of cell edges. The B- and C-grids are the ones that are generally used in existing ocean models.

Because of the staggered nature of the B- and C-grids, some spatial averaging is needed in order to implement the pressure gradient and Coriolis terms, respectively, and this averaging allows the possibility of various kinds of grid noise. Also, despite their relative superiority to other rectangular grid arrangements, the B- and C-grids can still allow substantial inaccuracy in the propagation of gravity waves and Rossby waves, depending on the grid resolution.

6.2 Unstructured Grids

Compared to regular rectangular grids, unstructured grids can provide a better fit to the complicated geography of the ocean's boundary. Also, some regions of the ocean, such as the western boundary currents, are more active than other regions, so

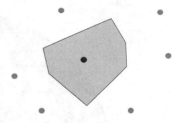

Fig. 9 Voronoi grid. The dots represent grid generators, and the shaded region is a grid cell

variable resolution within the fluid domain may be desirable. However, in the field of ocean circulation modeling, unstructured variable-resolution grids do not seem to be as firmly established as in other areas of computational science. A survey of some issues involved in three-dimensional ocean modeling on unstructured grids is given by Pain et al. [8].

For numerical methods on such grids, one possibility is to use finite element methods, either of the continuous Galerkin or discontinuous Galerkin variety. An example of the former is the finite-element shallow-water model of White et al. [11]. Discontinuous Galerkin (DG) methods have been used with the shallow water equations to model storm surges in localized regions, e.g., Dawson et al. [3]. The high locality of DG methods is an appealing feature, given the massive parallelism that is needed for large-scale ocean simulations.

Another option, which represents a major break from tradition, is to employ a Voronoi grid. See, e.g., Ringler et al. [9]. With such a grid, the starting point is to choose a set of points that serve as *grid generators*. (Here, the term "grid generator" has a different meaning from that used in Sect. 3.2.) For each grid generator, the corresponding grid cell is the set of points that are closer to that generator than to any other generator; see Fig. 9. If the grid generators are arranged in a rectangular array on a plane, then the grid cells are rectangles. On a spheroid, the vertices in a uniform triangulation could be used as grid generators; in that case the resulting grid cells are hexagons or pentagons, mostly the former, and the triangles can be regarded as a dual grid. More generally, the grid generators can be distributed arbitrarily, so as to produce a variable-resolution grid for which the grid cells are convex polygons. When partial differential equations are discretized on a Voronoi grid, divergence and curl are computed with a discrete vector calculus that uses line integrals around the boundaries of grid cells or dual cells. The authors of [9] are presently developing models of ocean circulation and atmospheric circulation with Voronoi grids as the method of spatial discretization.

7   Summary

This paper describes some physical and computational issues that are encountered when the circulation of the ocean is modeled numerically. These include multiple scales, the choice of vertical coordinate, properties of the governing equations, and time and space discretizations. Operational ocean models represent large-scale efforts by many people, and numerous models are already at a high state of development. However, this paper describes some areas of further work that could involve mathematical scientists, including time-stepping, hybrid vertical coordinate, variable-resolution horizontal grids, and methods for spatial discretization on such grids.

Acknowledgements I thank Rainer Bleck and Todd Ringler for useful discussions on matters related to the contents of this paper.

References

1. R. BLECK, *An oceanic general circulation model framed in hybrid isopycnic-Cartesian coordinates*, Ocean Modelling, 4 (2002), pp. 55–88.
2. R. BLECK, S. BENJAMIN, J. LEE, AND A. E. MACDONALD, *On the use of an adaptive, hybrid-isentropic vertical coordinate in global atmospheric modeling*, Monthly Weather Review, 138 (2010), pp. 2188–2210.
3. C. DAWSON, E. J. KUBATKO, J. J. WESTERINK, C. TRAHAN, C. MIRABITO, C. MICHOSKI, AND N. PANDA, *Discontinuous Galerkin methods for modeling hurricane storm surge*, Advances in Water Resources, 34 (2011), pp. 1165–1176.
4. S. M. GRIFFIES, *Fundamentals of Ocean Climate Models*, Princeton University Press, Princeton, N.J., 2004.
5. R. L. HIGDON, *A two-level time-stepping method for layered ocean circulation models: further development and testing*, J. Comput. Phys., 206 (2005), pp. 463–504.
6. R. L. HIGDON, *Numerical modelling of ocean circulation*, Acta Numerica, 15 (2006), pp. 385–470.
7. D. A. KNOLL AND D. E. KEYES, *Jacobian-free Newton–Krylov methods: a survey of approaches and applications*, J. Comput. Phys., 193 (2004), pp. 357–397.
8. C. C. PAIN, M. D. PIGGOTT, A. J. H. GODDARD, F. FANG, G. J. GORMAN, D. P. MARSHALL, M. D. EATON, P. W. POWER, AND C. R. E. DE OLIVEIRA, *Three-dimensional unstructured mesh ocean modelling*, Ocean Modelling, 10 (2005), pp. 5–33.
9. T. D. RINGLER, J. THUBURN, J. B. KLEMP, AND W. C. SKAMAROCK, *A unified approach to energy conservation and potential vorticity dynamics for arbitrarily structured C-grids*, J. Comput. Phys., 229 (2010), pp. 3065–3090.
10. A. F. SHCHEPETKIN AND J. C. MCWILLIAMS, *The regional oceanic modeling system (ROMS): a split-explicit, free-surface, topography-following-coordinate oceanic model*, Ocean Modelling, 9 (2005), pp. 347–404.
11. L. WHITE, E. DELEERSNIJDER, AND V. LEGAT, *A three-dimensional unstructured mesh finite element shallow-water model, with application to the flows around an island and in a wind-driven, elongated basin*, Ocean Modelling, 22 (2008), pp. 26–47.

Modeling Hazardous, Free-Surface Geophysical Flows with Depth-Averaged Hyperbolic Systems and Adaptive Numerical Methods

D.L. George

Abstract The mathematical modeling and numerical simulation of gravity-driven, free-surface geophysical flows—such as tsunamis, water floods, and debris flows—are described. These shallow flows are often modeled with two-dimensional depth-averaged equations that possess similar mathematical structure: they are usually nonconservative hyperbolic systems with source terms. Some numerical challenges presented by these systems, particularly with regard to features common to depth-averaged models for flow over topography, are highlighted. The open-source software package GEOCLAW incorporates numerical algorithms designed to tackle many of the common difficulties presented by depth-averaged models, particularly for large multiscale problems featuring inundation influenced by topography. Some simulation results for tsunamis, floods, and debris flows highlight the capabilities of GEOCLAW.

1 Introduction

A wide variety of regularly occurring natural hazards can be classified as shallow, gravity-driven, free-surface geophysical flows. This class of flows includes tsunami propagation and inundation; coastal storm surges; riverine and overland flooding due to rainfall or dam and levee failures; and finally a subclass of flows involving variable granular-fluid mixtures such as mudflows and debris flows. The impact of these phenomena continues to grow due to expanding population and development coupled with anthropogenic alteration of landscapes and waterways, which may be further exacerbated by climate change because of changing rainfall patterns and vegetation. Therefore, there is an increasing interest in modeling and simulating

D.L. George (✉)
U.S. Geological Survey, Vancouver, WA, USA
e-mail: dgeorge@usgs.gov

C. Dawson and M. Gerritsen (eds.), *Computational Challenges in the Geosciences*, The IMA Volumes in Mathematics and its Applications 156, DOI 10.1007/978-1-4614-7434-0_2,
© Springer Science+Business Media New York 2013

such flows for the purposes of hazard assessment, land development and management, and infrastructure placement and design. More generally, simulations of these kinds of flows are becoming an integral part of the emerging field of sustainability science and engineering (see, e.g., [1]).

From a modeling and simulation perspective, these types of flows present similar challenges. Typically they are modeled with depth-averaged equations for conservation of mass and momentum for single or multiple constituent phases or layers. The resulting coupled partial differential equations (PDEs) are usually nonlinear hyperbolic systems. Aside from the usual well-known numerical challenges posed by hyperbolic systems (e.g., shocks and nonunique discontinuous weak solutions), depth-averaged equations for free-surface flows present a number of unique challenges. For instance, the flow usually advances over, or is routed by, complex topography, which introduces source terms in the momentum equations that can be discontinuous or singular. Additionally, the advancing inundation front represents a moving domain boundary that must be tracked or captured accurately and stably. A more subtle difficulty concerns numerical steady-state preservation: the dynamics of these free-surface flows are often small perturbations to an underlying balanced steady state where the flux gradients and source terms are large and counterbalanced. Finally, these flows have a large spatial extent and contain features with small length scales. Therefore, they are extremely multiscale in nature. For example, a tsunami wave might cross an entire ocean, yet coastal inundation can depend on meter-scale topographical features thousands of kilometers away from the source. A flood or debris flow might inundate an irregular region that is bounded by thousands of square kilometers, yet accurately modeling the flow can require sub-meter resolution.

For many applications involving water flow, such as tsunamis, river flow, and overland flooding, the use of simple depth-averaged models such as the shallow water equations has been well established for decades (see, e.g., [2,3]). For problems in which there may be more vertical variation in the flow-velocity profile, for example storm surges [4], some researchers have adopted multilayer depth-averaged models (see, e.g., [5]). These multilayer equations are also nonlinear hyperbolic systems, but they pose additional mathematical challenges, such as loss of hyperbolicity when flow velocities in different layers diverge beyond some threshold [5]. Nevertheless, it is possible to develop stable numerical schemes that avoid instabilities due to loss of hyperbolicity (see, e.g., [6,7]).

Depth-averaged models for free-surface flows involving mixtures of particles and fluid are much less well established, and simply developing the mathematical governing equations remains a significant challenge. Most models are based on depth-averaging two-phase flow with a theoretical approximation for the complicated stress that arises due to intergranular and granular-fluid interactions. The resulting equations are often nonlinear hyperbolic systems similar in mathematical structure to the shallow water equations, albeit with multiple phases and more complicated source terms (see, e.g., [8,9]). Determining the suitability of, and improving, such models is an active area of research.

The aim of this paper is to introduce and survey some of the common challenges inherent in modeling free-surface flows that move over topography. The common mathematical features of several depth-averaged equations are summarized. Additionally, numerical techniques developed to overcome these challenges are described. The full detail of the numerical techniques are left to the references—the aim here is to provide an introduction and survey of the methods implemented in GEOCLAW that are specifically relevant to these applications. The paper is organized as follows: in Sect. 2 mathematical models are introduced; in Sect. 3 some numerical challenges, methods, and software are described; and finally, in Sect. 4 some specific applications and a sample of simulation results are presented.

2 Mathematical Models

In this section hyperbolic systems are first described in general form before some depth-averaged models that belong to this class of equations are surveyed.

2.1 Hyperbolic Conservation Laws and Nonconservative Balance Laws

Hyperbolic systems of PDEs are commonly derived as integral conservation laws for mass, momentum, energy, or other conserved quantities. For a vector $q(x,t) \in \mathcal{R} \subseteq \mathbb{R}^m$ of m conserved quantities, a hyperbolic conservation law (in one spatial dimension) is of the form

$$\partial_t q + \partial_x f(q) = 0, \tag{2.1}$$

where $f : \mathbb{R}^m \to \mathbb{R}^m$ is a flux function and $\mathcal{R}$ represents physically admissible solution states. In order to satisfy the property of hyperbolicity, by definition the flux Jacobian $f'(q) \in \mathbb{R}^{m \times m}$ must be diagonalizable with real eigenvalues for all $q \in \mathcal{R}$. If $f'(q)$ satisfies these requirements only for some subset of states $q \in \mathcal{R}' \subsetneq \mathcal{R}$, the system is not strictly hyperbolic, and loss of hyperbolicity when $q \notin \mathcal{R}'$ results in an unstable or ill-posed system (see, e.g., [10]). Aside from this, even strictly hyperbolic systems present some well-known difficulties. For instance, nonlinear systems such as Eq. (2.1) lead to wave steepening and discontinuities, yet, due to the integral form from which Eq. (2.1) is derived, discontinuous weak solutions can be determined through Rankine–Hugoniot jump conditions based on boundary fluxes. However, this opens the possibility of nonunique weak solutions, which necessitates additional physically based admissibility conditions that are extraneous to Eq. (2.1). These are often called entropy conditions, analogous to the entropy requirements for the Euler equations of gas dynamics.

Most traditional finite difference schemes for PDEs do not converge to generalized weak solutions of Eq. (2.1), and in fact they are numerically unstable in the presence of discontinuities or sharp gradients. This has prompted the development of specialized shock-capturing schemes that can converge stably to admissible weak solutions, provided that some admissibility criteria are known and used in the design of the numerical scheme. Many of these methods are Godunov-type methods (i.e., methods making use of Riemann solvers for the numerical update). The high-resolution wave-propagation methods employed in GEOCLAW, described briefly in Sect. 3, are one example of Godunov-type methods.

For many applications, the solution $q(x,t)$ is not conserved but obeys a nonconservative balance law,

$$\partial_t q + \mathcal{A}(q)\partial_x q = \psi(q,x), \tag{2.2}$$

where $\psi : \mathbb{R}^m \times \mathbb{R} \to \mathbb{R}^m$ is a source term and $\mathcal{A} : \mathbb{R}^m \to \mathbb{R}^{m \times m}$ is some matrix-valued function. Similar to a conservation law, Eq. (2.2) is hyperbolic if $\mathcal{A}(q)$ is diagonalizable with real eigenvalues. The quasilinear form (2.2) is a more general form of a hyperbolic system: it includes Eq. (2.1) in the special case in which $\mathcal{A}(q) = f'(q)$ and $\psi(q) \equiv 0$. More generally, the term $\mathcal{A}(q)\partial_x q$ could be a nonconservative product (i.e., if $\mathcal{A}(q)$ is not a full derivative of a flux). Systems with nonconservative products present additional mathematical challenges. Because there is no flux function to provide Rankine–Hugoniot jump conditions for discontinuities, determining weak solutions near a discontinuity can be difficult (see, e.g., [11–14]), but finding such solutions can be advantageous in the design of numerical schemes (see, e.g., [13, 15]).

In two dimensions (2D), a hyperbolic system takes the general form

$$\partial_t q + \mathcal{A}(q)\partial_x q + \mathcal{B}(q)\partial_y q = \psi(q,x,y), \tag{2.3}$$

where similar to Eq. (2.2), $q(x,y) \in \mathbb{R}^m$, $\psi : \mathbb{R}^m \times \mathbb{R}^2 \to \mathbb{R}^m$, and $\mathcal{A},\mathcal{B} : \mathbb{R}^m \to \mathbb{R}^{m \times m}$. The depth-averaged models described in Sect. 2.2 are of this general form.

2.2 Depth-Averaged Equations

For many geophysical surface flows, it is common to apply the shallow approximation, that is, the assumption that the flow depth $h(x,y,t)$ is small relative to the characteristic horizontal length scale H (see Fig. 1). This assumption justifies integrating three-dimensional (3D) governing equations through the depth and applying boundary conditions at the basal and free surfaces, resulting in a 2D system that is more tractable computationally. This procedure also incorporates the free-surface boundary conditions directly in the governing PDEs, so that depth becomes a solution variable. In depth-averaged models the vertical variation in the flow field (of a particular layer) is either ignored or assumed to have a given profile. The models described in this section are for a fluid or continuum mixture with a free

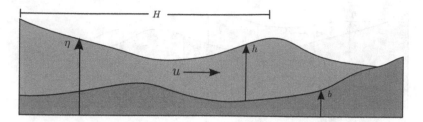

Fig. 1 Vertical cross section of a single-layer depth-averaged flow. The fluid (*light shade*) with depth h flows over fixed bottom topography b (*dark shade*). The shallowness parameter $\varepsilon = h/H$ is assumed to be small

surface at $z = \eta(x,y,t)$, flowing over fixed basal topography at $z = b(x,y)$. The total flow depth is then $h(x,y,t) = \eta(x,y,t) - b(x,y)$. The horizontal flow velocities in the x and y directions will be denoted by $u(x,y,t)$ and $v(x,y,t)$ respectively. Multilayer and multiphase flows will be denoted with subscripts for different layers or phases.

2.2.1 The Shallow Water Equations

The simplest depth-averaged equations are the well-known shallow water equations,

$$\partial_t h + \partial_x(hu) + \partial_y(hv) = 0, \tag{2.4a}$$

$$\partial_t(hu) + \partial_x(hu^2 + \tfrac{1}{2}gh^2) + \partial_y(huv) = -gh\partial_x b + S_x, \tag{2.4b}$$

$$\partial_t(hv) + \partial_x(huv) + \partial_y(hv^2 + \tfrac{1}{2}gh^2) = -gh\partial_y b + S_y, \tag{2.4c}$$

where g is the gravitational acceleration constant, and $hu(x,y,t)$ and $hv(x,y,t)$ are the depth-averaged momenta in the x and y directions, respectively. The terms $S_x(q,x,y)$ and $S_y(q,x,y)$ are basal friction terms obtained from an empirical friction law, such as the Chezy or Manning formula (see, e.g., [16, 17]).

The shallow water equations are derived (see, e.g., [2]) under the assumption that the fluid pressure is vertically hydrostatic, which can be justified if the shallowness parameter $\varepsilon = h/H$ is very small. In this case deviations from hydrostatic pressure can be formally shown to be $\mathcal{O}(\varepsilon^2)$ (see, e.g., [2]). These equations are commonly used, however, for problems in which this assumption is violated in some regions. Nevertheless, this system of equations has been repeatedly shown to perform very well for a wide variety of applications, at least for predicting general inundation patterns (see, e.g., [17, 18]). For problems in which the flow exhibits large vertical accelerations, and hence nonhydrostatic pressure that could significantly influence the dynamics, some authors have included nonhydrostatic pressure corrections to the equations (see, e.g., [19–23]). Inclusion of such correction terms introduces third-order dispersive terms, altering the mathematical nature of the system so that it is no longer of the form (2.3) (see, e.g., [19, 22]). This could potentially be dealt with numerically by splitting the hyperbolic and dispersive terms into separate integration steps.

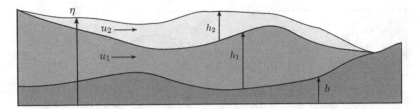

Fig. 2 Vertical cross section of a two-layer depth-averaged flow. The fluid (*lightest shade*) with depth h_2 flows over another fluid layer with depth h_1 and fixed bottom topography b (*dark shade*)

2.2.2 Multilayer Models

For problems in which there may be a large degree of vertical variation in the horizontal velocity profiles or fluid densities, researchers have used multilayer shallow water equations (Fig. 2). The solution $q(x,y,t)$ for an N-layer system takes the form

$$q = (\rho_1 h_1, \ldots, \rho_N h_N, hu_1, \ldots, hu_N, hv_1, \ldots, hv_N)^{\mathrm{T}}, \qquad (2.5)$$

where ρ_n is the constant density, $h_n(x,y,t)$ is the depth, and $hu_n(x,y,t)$ and $hv_n(x,y,t)$ are the depth-averaged momenta of the nth layer. The exact form of the governing system depends on layer-coupling assumptions, but they are usually of the form (2.3) (see, e.g., [5–7, 24]). Multilayer models are often used for ocean dynamics problems in which a relatively shallow, more dynamic layer overlies denser and deeper layers in the water column. For instance, multilayer models have been used for storm surge generation and propagation [5]. One difficulty encountered with multilayer models is the loss of hyperbolicity due to the appearance of complex eigenvalues when layer velocities diverge significantly, i.e., when $\|u_n - u_{n-1}\| > M_n$ at some location for $n = 2, \ldots, N$. The values M_n depend on the density ratio and depths of the adjacent layers. However, this problem can be overcome numerically with various specialized methods that prevent the loss of hyperbolicity (see, e.g., [6, 7, 25]). Interestingly, the problem can be avoided altogether if mass exchange is allowed between the layers [24].

2.2.3 Granular-Fluid Flows

Gravity-driven mass flows such as debris flows are commonly composed of saturated mixtures of solid particles, interstitial fluid, and suspended fine sediment (see, e.g., [26, 27] for an overview). The most notable difficulty in developing models for these flows is representing the complicated stress with some physical fidelity, while maintaining mathematical and computational tractability. The use of depth-averaged models for these mass flows was pioneered by Savage and Hutter [28, 29], who developed a model similar to the shallow water equations, yet accounted for solid stress using a solid-pressure model and Coulomb friction. In their model the fluid phase is ignored altogether, making it strictly appropriate for dry granular avalanches.

In order to model fluid-saturated flows, some researchers have treated granular flows as a single-phase continuum with a non-Newtonian rheology, such as a Bingham or viscoplastic medium (see, e.g., [30]). Iverson and Denlinger [31, 32] in contrast utilized a Coulomb-based Savage–Hutter-type model for granular flow, but included terms to explicitly account for the solid–fluid stresses in a mixture treated as a single phase. More recently, researchers (see, e.g., [8, 33–35]) have developed multiphase continuum mixture models, usually with one phase for the solid particles and another for interstitial fluid with suspended fine particles. These multiphase models allow for a more accurate accounting of the stress arising from solid–solid and solid–fluid interactions, but are inherently more complex. Typically the solid stress terms are based on Coulomb theory, and solid–fluid interactions are modeled with a drag term (arising from phase velocity differences) and a buoyancy term (arising from interstitial fluid pressure).

With a typical depth-averaged two-phase model, the solution vector might take the form

$$q = (\alpha_s h, \alpha_f h, hu_s, hv_s, hu_f, hv_f)^{\mathrm{T}}, \qquad (2.6)$$

where subscripts s, f indicate the solid or fluid phase respectively, and $\alpha_{s,f}$ are the volume fractions of the two phases. Various changes of variables with functions of the volume fractions and depth are of course possible as well, e.g., [33]. Because of the nature of the stress laws used, the governing PDEs are usually hyperbolic systems of the form (2.3) (e.g., [8, 33–35]). These models present the same mathematical difficulty as most multilayer shallow water models, in that complex eigenvalues can appear when there are large differences in phase velocities. Additionally, the eigenstructure cannot generally be determined analytically (see, e.g., [8, 34]).

Although multiphase models include buoyancy terms arising from pore-fluid pressure, accurately representing that pressure remains a significant challenge. The dynamics of granular flows are heavily influenced by the expansion and contraction of the pore space during solid grain motion, which creates a feedback loop between fluid pressure and pressure-mediated solid mobility (see, e.g., [27, 36]). In most models this effect is neglected and the fluid pressure evolves independently from the solid volume fraction (see, e.g., [8, 32]). A recently developed model [9] accounts for this feedback by using the concept of granular dilatancy: originally developed in the field of quasi-static soil mechanics (see, e.g., [37]). Although the details of the model are beyond the scope of this paper, in essence the pore-fluid pressure is coupled to the rate of expansion or contraction of the solid volume fraction, which evolves due to shearing, allowing feedback and coevolution of these variables (see [9, 36] for details). The variables modeled are

$$q = (h, hu, hv, \alpha_s, p)^{\mathrm{T}}, \qquad (2.7)$$

where p is the pore-fluid pressure. Although independent phase velocity fields are not tracked, the model is two-phase in the sense that the volume fractions and fluid pressure are retained and coevolve. The model has the added benefit of being strictly hyperbolic with a simple analytically known eigenstructure. Because pressure is not a conserved quantity, the system has nonconservative products that must be handled numerically (described briefly in Sect. 3).

2.2.4 Hybrid Models

Some geophysical applications contain multiple interacting flow phenomena, each of which can be described by a distinct depth-averaged model. Hybrid models formed by coupling together individual depth-averaged equations are sometimes used for these applications. For instance, sediment transport problems in which water flows over loose sediment are modeled with the shallow water equations coupled with a hyperbolic advection equation for the entrained sediment, resulting in an extended system of the form (2.3) (see, e.g., [38, 39]). If the bed load transport is modeled in addition to suspended sediment, a two-layer set of equations can be used, with the lower layer representing the flowing granular material below the water and suspended load. Typically very simple models are used for the lower layer, because its dynamics are mostly due to the motion of the overlying water, rather than being independently gravity-driven (see, e.g., [38]). For problems involving submarine landslides and turbidity currents, the granular flows are gravity-driven and they have more influence on the motion of the overlying water column than does bed load. For these problems it can be advantageous to use multilayer models with the lower layer being governed by a granular-fluid model, such as those introduced above, and the water being modeled with the shallow water equations (see, e.g., [40]). Additionally, there is a growing interest in submarine-landslide-generated tsunamis (e.g., [41]), which could potentially be described by similar multilayer models.

3 Numerical Techniques and Challenges

In this section the numerical algorithms implemented in GEOCLAW [42] are briefly described in general terms, followed by some examples for depth-averaged flows. An exhaustive and detailed description of the algorithms is far beyond the scope and focus of this paper. Rather, the central aim of this section is to highlight some of the challenges and techniques specific to simulating free-surface flows over topography.

3.1 Finite-Volume Wave-Propagation Methods

GEOCLAW is based on two-dimensional finite-volume methods. The grids are logically rectangular, but various mappings, such as latitude–longitude mappings, are supported for modeling global-scale problems on the sphere (see Fig. 3).

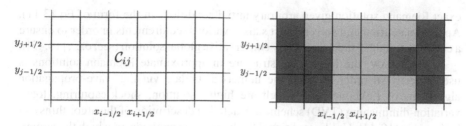

Fig. 3 *Left*: map view of a logically rectangular finite-volume grid in 2D. The x and y coordinates might represent Cartesian or latitude–longitude coordinates. *Right*: finite-volume grid for depth-averaged flows. The edge of the flow is captured as the boundary between the wet cells (indicated in a *light shade*) and the dry cells (indicated in a *darker shade*)

The numerical solution at a given time t^n is defined as piecewise constant,

$$Q_{ij}^n \approx \frac{1}{\lambda(C_{ij})} \int_{C_{ij}} q(x,y,t^n)\,dx\,dy, \tag{3.8}$$

where $\lambda(C_{ij})$ is the area of the grid cell $C_{ij} = [x_{i-1/2},x_{i+1/2}] \times [y_{j-1/2},y_{j+1/2}]$. In the following sections, capital letters and subscripts are used to denote the components, or functions of the components, of the numerical solution Q_{ij}^n (e.g., $U_{ij}^n = HU_{ij}^n/H_{ij}^n$). Each grid cell also has a topographic value,

$$B_{ij} = \frac{1}{\lambda(C_{ij})} \int_{C_{ij}} b(x,y)\,dx\,dy, \tag{3.9}$$

where $b(x,y)$ is determined from external data such as a digital elevation model (DEM). The numerical update $Q_{ij}^n \to Q_{ij}^{n+1}$ requires the solution of one-dimensional Riemann problems at each of the four grid-cell interfaces. For example, in solving a system defined by Eq. (2.3), at the interface between cells $C_{i-1,j}$ and C_{ij}, the system

$$\partial_t q + A(q)\partial_x q = \psi(q,x,y), \quad \text{for} \quad t^n \le t \le t^{n+1}, \tag{3.10}$$

is analytically solved with initial conditions,

$$q(x,y,t^n) = \begin{cases} Q_{i-1,j}^n, & \text{if} \quad x \le x_{i-1/2}, \\ Q_{ij}^n, & \text{if} \quad x > x_{i-1/2}. \end{cases} \tag{3.11}$$

An analogous Riemann problem is solved for the interface between cells $C_{i+1,j}$ and C_{ij}. For the interfaces between cells C_{ij} and $C_{i,j\pm1}$, the system

$$\partial_t q + B(q)\partial_y q = \psi(q,x,y), \quad \text{for} \quad t^n \le t \le t^{n+1}, \tag{3.12}$$

is solved with analogous initial conditions. Typically an approximate Riemann solver, developed for a given set of PDEs, provides an approximation to the exact Riemann solution given arbitrary initial conditions in the form of Eq. (3.11). Approximate Riemann solvers must satisfy various requirements in order to ensure a consistent, stable, and numerically conservative global solution (see, e.g., [10,43]).

In GEOCLAW, the local wave structure in approximate Riemann solutions is used directly in order to build the numerical update via the wave-propagation algorithms of LeVeque [44], which are high-resolution, shock-capturing, total-variation-diminishing (TVD) schemes. A detailed description of the algorithms can be found in [10,44]. Unlike traditional Godunov-type methods, in which the numerical update is based on cell-interface fluxes determined from the Riemann solutions, the wave-propagation algorithms are applicable to systems with nonconservative products where there is no flux function.

Note that for an advancing or retreating flow, the location of the moving flow front is determined by the stair-step boundary between wet and dry cells at any given time step (see Fig. 3). While some inundation algorithms explicitly track the boundary by solving a separate set of kinematic equations for the edge trajectory (see, e.g., [45]), the algorithms in GEOCLAW capture the boundary by simply solving the Riemann problem between wet and dry cells, allowing dry cells to fill or wet cells to drain. This requires specialized Riemann solvers that carefully treat the source term resulting from variable topography, introduced in the next section.

3.2   Riemann Problems for Flow over Topography

Gradients in topography appear in the source terms of all of the models described in Sect. 2.2. Numerically including these terms in a satisfactory manner is surprisingly more difficult than might be expected, and the topic has generated a considerable amount of research (see, e.g., [15,46,47]). In GEOCLAW the topography is dealt with directly in the Riemann solver. To illuminate the challenges, consider a Riemann problem for the shallow water equations (2.4), as depicted in Fig. 4. The figure shows a step in topography, surface elevation, and fluid depth. It is physically intuitive that the waves in the Riemann solution arise due to deviations from the stationary steady state: $\eta_{i-1,j}^n = \eta_{ij}^n$, $U_{i-1,j}^n = U_{ij}^n = 0$, where η represents the free surface of the numerical solution, $\eta_{ij}^n = H_{ij}^n + B_{ij}^n$. Mathematically, the steady state exists due to a balance between nontrivial convective terms on the left-hand side of Eq. (2.4) (arising from depth gradients) and the nontrivial source term on the right (arising from topography gradients). Written compactly in the general form (3.10), the steady state satisfies

$$A(q)\partial_x q = \psi(q,x,y). \tag{3.13}$$

(Note that for the Riemann problem between $C_{i-1,j}$ and C_{ij}, variation in y can be ignored.) Traditionally, Riemann solvers (e.g., [48, 49]) are designed for the homogeneous system

$$\partial_t q + A(q)\partial_x q = 0, \tag{3.14}$$

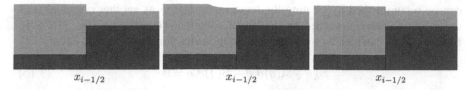

$$x_{i-1/2} \qquad\qquad\qquad x_{i-1/2} \qquad\qquad\qquad x_{i-1/2}$$

Fig. 4 Vertical cross-section of a one-dimensional Riemann problem for a typical depth-averaged free-surface flow, showing the fluid (*light*) and topography (*dark*) in cells $C_{i-1,j}$ and C_{ij}. *Left*: initial solution at t^n. *Middle*: the Riemann solution at $t > t^n$ consists of waves (smooth or discontinuous) that propagate away from the initial discontinuity at $x_{i-1/2}$, and represent deviations to the steady state. *Right*: the re-averaged solution at t^{n+1}

neglecting the source term. The source term is then accounted for in a separate integration step (from t^n to t^{n+1}) for the system of ordinary differential equations

$$\partial_t q = \psi(q, x, y). \tag{3.15}$$

These fractional step or split methods fail to preserve balanced steady states (i.e., nontrivial steady states) at practical grid resolutions, particularly when the balanced terms are large in magnitude. This should not be surprising, because the numerical updates based on Eqs. (3.14) and (3.15) must precisely cancel, yet might be large individually. Consequently, these methods also fail to correctly resolve small perturbations to such steady states, which characterize the dynamics of many shallow, free-surface flows [50]. Deep-ocean tsunami modeling provides a clear example: tsunamis represent a tiny perturbation to an underlying balanced steady state for still water over the variable seafloor.

3.2.1 Well-Balanced Methods

The challenges involving balanced steady states have motivated the development of well-balanced schemes, which are designed to incorporate the effect of the source term directly into the Riemann solution. Although simple in concept, this can be challenging to accomplish accurately. Again consider the Riemann problem for the shallow water equations depicted in Fig. 4. The source term due to the topographic gradient $\partial_x b$ is concentrated at the cell interface and is weakly defined as a delta function. However, note that h also appears in the source term of Eq. (2.4) and is discontinuous at the cell interface, creating a problem of overlapping distributions (see, e.g., [11, 12, 51]). The correct magnitude for the source term is therefore unclear, yet one wishes to exactly preserve (to machine precision) steady-state discontinuous weak solutions.

The correct solution to this problem can be illuminated by transforming the shallow water equations (neglecting friction) into a homogeneous system, which can be accomplished by treating the topography $b(x, y)$ as another component of the solution vector, governed by the trivial equation $\partial_t b = 0$ (see, e.g., [15, 52]). Equation (2.4) can then be written as a homogeneous nonconservative system,

$$\partial_t \tilde{q} + \tilde{A}(\tilde{q})\partial_x \tilde{q} + \tilde{B}(\tilde{q})\partial_y \tilde{q} = 0, \tag{3.16}$$

where $\tilde{q} := (h, hu, hv, b)^{\mathrm{T}}$ and $\tilde{A}, \tilde{B} : \mathbb{R}^4 \to \mathbb{R}^{4 \times 4}$. With the equations in this form, the jump discontinuity that remains at the cell interface for $t \geq t^n$ is a stationary wave in the Riemann solution variables $\tilde{q}$, which must be determined by jump conditions. To see how these jump conditions can be derived, consider a smooth steady-state solution in 1D $\tilde{q}(x, \cdot, \cdot)$, on some interval $x_0 \leq x \leq x_1$, with varying topography. From Eq. (3.16), the solution must satisfy

$$\tilde{A}(\tilde{q})\partial_x \tilde{q} \equiv 0. \tag{3.17}$$

Consider this solution to be a curve in state space, parameterized by $x : \tilde{Q}(x) := \tilde{q}(x, \cdot, \cdot)$. From Eq. (3.17) it can be seen that at any point $x_p \in [x_0, x_1], \partial_x \tilde{q}(x_p, \cdot, \cdot) = \tilde{Q}'(x_p)$ is an eigenvector $r_0(\tilde{q}(x_p))$ of $\tilde{A}(\tilde{q}(x_p))$ associated with a zero eigenvalue, $\lambda_0(\tilde{q}(x_p)) = 0$. The solution in phase space is therefore an integral curve of the vector field $r_0(\tilde{q})$:

$$\tilde{Q}'(x) = \alpha(x)r_0(\tilde{q}(x)), \quad x_0 \leq x \leq x_1, \tag{3.18}$$

where $\alpha(x)$ is a scalar proportionality constant. The solution varies smoothly along the curve in state space as x varies. The relevant discontinuous steady-state solution with a single jump in topography is therefore a jump discontinuity between two states lying on the same integral curve of $r_0(\tilde{q})$, which connects $\tilde{q}(x_0)$ and $\tilde{q}(x_1)$. This jump discontinuity provides the correct jump conditions for the two states adjacent to $x_{i-1/2}$, such as shown in the middle panel of Fig. 4 (see, e.g., [15, 53]). Well-balanced methods are designed to approximate this stationary wave as accurately as possible, though usually without determining the exact Riemann solution (see, e.g., [13–15, 50, 54–57]). Consequently, most well-balanced methods preserve some steady-state solutions (e.g., motionless water) but not all (e.g., steady flow). The well-balanced Riemann solver used in GEOCLAW preserves all one-dimensional steady states and the motionless steady state in 2D. However, exactly preserving general flowing steady states in 2D remains a challenging problem [15].

3.2.2 Wet–Dry States

Another challenge confronted in modeling free-surface flows that inundate dry land is accurately resolving the Riemann problem between a wet and dry cell. This is sometimes referred to as the dry-state problem or the filling and draining problem. Conventional approximate Riemann solvers are prone to producing negative depths for these and other very shallow Riemann problems. Simply resetting the negative

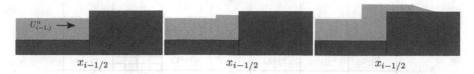

Fig. 5 The wet–dry Riemann problem with a step in topography. *Left*: the initial left state has fluid with positive velocity. *Middle*: the Riemann solution when the velocity is insufficient to inundate the dry cell. The solution consists of reflected waves. *Right*: the Riemann solution with inundation

depth to zero violates conservation and redistributes mass inaccurately in the inundation zone, where accuracy is typically the most desirable [17,58]. A number of researchers have therefore developed so-called positivity-preserving or depth-positive-semidefinite approximate Riemann solvers that avoid this problem (see, e.g., [15,59,60]). These solvers accurately capture the draining of a wet cell.

Even when maintaining depth-positivity, solving the Riemann problem between wet and dry states when there is a step in topography remains a challenging problem, and it is well known that a standard discretization of $b(x,y)$ for these problems will produce spurious results for the advancing inundation edge [17, 61]. This becomes clear when one considers the inundation Riemann problems shown in Fig. 5. Consider the initial data in the wet cell $\mathcal{C}_{i-1,j}$, which has a surface elevation $\eta_{i-1,j}^n < B_{ij}$ and a positive velocity $U_{i-1,j}^n$. It is clear that there are two distinct possibilities in the Riemann solution: either the fluid will be completely reflected by the step in topography, or it will inundate the dry cell to the right. The correct solution depends on the initial fluid velocity. However, the Riemann solution also depends on the source term, which is undefined at t^n for $x \geq x_{i-1/2}$. This ambiguity is physically intuitive, since in the case in which waves are entirely reflected, a value of B_{ij} higher than the fluid level has no influence on the solution. In GEOCLAW, before the wet–dry Riemann problem is actually solved, a preliminary test problem is conducted with a wall boundary condition (i.e., solution values symmetric to $Q_{i-1,j}^n$ are used in $\mathcal{C}_{ij}$). The solution of this test problem determines whether $U_{i-1,j}^n$ is of sufficient magnitude to overtop the step in topography. If it is not, the reflected waves are retained for the update $Q_{i-1,j}^n \to Q_{i-1,j}^{n+1}$, and $\mathcal{C}_{ij}$ remains dry. If instead the waves reflected from the wall are higher than B_{ij}, then the original wet–dry Riemann problem is solved with a standard discretization of the source term. Although this ad hoc test determines whether inundation occurs, solving the inundation problem exactly when there is a step in topography is difficult (see, e.g., [62]). Because of the computational expense associated with determining the exact solution, the solution is approximated in GEOCLAW [15].

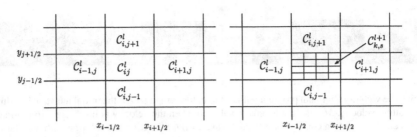

Fig. 6 Block-structured adaptive mesh refinement. A single level-l grid cell C_{ij}^l is refined to 16 level-$(l+1)$ cells, with refinement ratios $r_x^l = r_y^l = 4$

3.3 Adaptive Mesh Refinement

In order to accommodate the multiple spatial and temporal scales in geophysical problems, GEOCLAW utilizes block-structured, patch-based adaptive mesh refinement (AMR), designed for logically rectangular finite volume grids [63–65]. These algorithms allow multiple grid *levels*, $l = 1, \ldots, L$, with different resolutions. Spatial refinement ratios between grid levels are user-defined integers for each direction: r_x^l, r_y^l, for $l = 1, \ldots, L-1$. For instance, Fig. 6 depicts the refinement of a single level-l grid cell C_{ij}^l to a level-$(l+1)$ grid with $r_x^l = r_y^l = 4$. Typically, many adjacent grid cells of a given level-l grid would be refined to generate a single logically rectangular level-$(l+1)$ subgrid. The individual subgrids of a given level are arranged to form a patchwork that evolves with features in the solution. By using many levels of refinement, the diversity of spatial scales in a given problem can be accommodated.

The AMR algorithms used in GEOCLAW are based on methods developed for shock-capturing methods applicable to general hyperbolic conservation laws by Berger, Colella, and Oliger [63, 64]. These were later generalized to accommodate wave-propagation algorithms for conservative and nonconservative problems [65]. The algorithms are designed to interpolate and coarsen (*anterpolate*) the solution between coarse and fine grids upon refinement and de-refinement. The interpolation/anterpolation schemes must be designed to maintain properties of the underlying numerical methods, such as numerical conservation and TVD maintenance in the case of shock-capturing schemes. Special techniques must be used at subgrid boundaries in order to maintain conservation and prevent reflections. Details of these algorithms can be found in [63–65].

Free-surface environmental flows present unique challenges for AMR algorithms, due particularly to the time-varying domain, vanishing depths near flow edges, and need for well-balanced preservation of steady states. Standard interpolation/anterpolation schemes (e.g., [63–65]) fail to preserve steady states, destroying the underlying well-balanced properties of the numerical scheme, as well as introducing energy in the form of hydraulic head. Additionally, because of nonlinear topography, it is possible to generate finite momentum in cells with vanishing fluid depths near the flow front, introducing unbounded velocities and kinetic energy. To overcome these challenges, an extensive set of interpolation/anterpolation schemes tailored to free-surface flows over topography were developed and implemented in GEOCLAW [17, 66]. In summary, the schemes satisfy several properties simultaneously. Steady states are maintained numerically, at and away from the flow edge. Energy (hydraulic head) is not introduced upon interpolation, and velocity interpolation is TVD (i.e., new extrema in fluid velocities are not introduced). For free-surface flows, AMR interpolation/anterpolation can violate conservation of mass. For example, with a body of water near the shoreline, it is impossible to maintain conservation of mass and steady states simultaneously [66], due to the changing representation of the shoreline on grids of different resolutions. However, the AMR schemes in GEOCLAW conserve momentum when motionless mass is introduced near a shoreline, and if mass with nonzero momentum is removed upon coarsening, the momentum associated with that mass is removed as well [66]. In practice, lack of conservation near shorelines is not particularly problematic, since refinement ideally occurs just prior to the arrival of waves carrying momentum. The unique interpolation/anterpolation schemes in GEOCLAW confer stability of the AMR algorithms for free-surface flows over topography [66].

4 Applications and Some Simulation Results

4.1 Tsunami Modeling: The 2011 Tohoku-Oki Tsunami

Tsunamis are typically modeled with the shallow water equations (2.4), since the depth of the ocean is very small compared to typical wavelengths. Tsunamis are typically generated by the vertical displacement of the seafloor,

$$b(x,y) \rightarrow b(x,y) + \delta b(x,y,t),$$

which disturbs the idealized motionless steady-state ocean,

$$0 \equiv \eta(x,y,\cdot) \rightarrow \eta(x,y,t) + \delta b(x,y,t),$$

thereby generating waves. Seafloor motion can occur due to earthquake-generated fault motion or large submarine landslides. The former generation mechanism is commonly known to produce large teletsunamis, which propagate across entire ocean basins. The initial spatial scales of a teletsunami are related to those of the initial seafloor disturbance $\delta b(x,y,t)$, which has horizontal length scales typically on the order of 100 km with amplitudes on the order of 1 m or less [17]. The wave speeds of a tsunami are equivalent to the eigenvalues of $f'(q)$ for the shallow water equations (2.4), which are $u, u \pm \sqrt{gh}$. The dependence of the wave speed on water depth means that the leading edge of a wave slows in shallow coastal

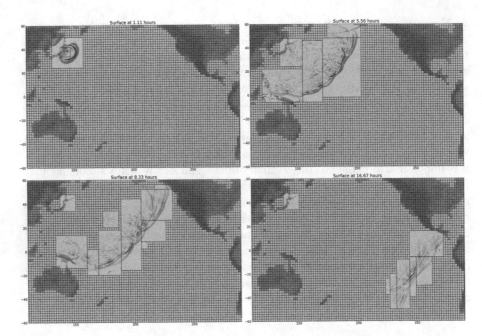

Fig. 7 A GEOCLAW simulation of the 11 March 2011 Tohoku-Oki tsunami. The steady-state ocean is maintained on extremely coarse level-1 grids, while level-2 grids (refined by 16^2) resolve the waves as they cross the ocean. Grid lines are omitted from level-2 grids for clarity. AMR provides a significant efficiency enhancement over static grids: this simulation required 27 min of computational time to model 24 h of real time (on a 64-bit Intel i7 quadcore desktop) (color figure online)

waters, which compresses the tsunami considerably and results in extreme energy focusing, giving rise to the tsunami's destructive potential. Upon inundation, the highly variable topographic features near the shore (seawalls, for example) can require meter-scale resolution for accurate modeling [17]. These disparate spatial scales imply that tsunamis are extremely multiscale in nature. Furthermore, because the waves propagate throughout the domain but are localized at any given time, the required grid resolution is highly spatially and temporally dependent, making AMR valuable if not essential for efficient modeling.

A typical refinement scenario for a GEOCLAW simulation is depicted in Figs. 7 and 8. In these simulations of the Tohoku-Oki tsunami of 2011, a single coarse level-1 grid with a resolution of 160 km was used for the simulation domain. The role of the coarse grid was to maintain the motionless steady state, while propagating waves were resolved on level-2 grids, refined by $r_x^1 = r_y^1 = 16$, resulting in a 10-km resolution. The simulation depicted in Fig. 7 used only these two levels.

Inundation modeling requires multiple (> 2) levels of grids, with high refinement ratios. Typically a region or coastline of interest is selected prior to the simulation. Figure 8 shows a GEOCLAW simulation for potential inundation in Hilo, Hawaii, caused by the Tohoku-Oki tsunami. Level-3 to level-5 grids were used to resolve

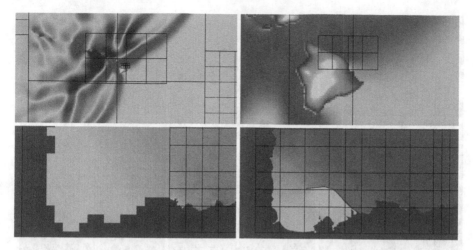

Fig. 8 A GeoClaw simulation showing inundation modeling, requiring extremely high refinement. *Upper panels*: as tsunami waves approach the northern coast of Hawaii, level 2–4 grids appear. *Lower panels*: inundation modeling of Hilo, Hawaii, on level-5 grids with ≈ 5 m resolution. The level-5 grids appear just as the waves approach the Hilo harbor. Grid lines are omitted from level 3–5 grids for clarity. AMR allows transoceanic propagation and local inundation to be modeled in single global-scale simulations (color figure online)

the waves as they compressed in the shallow coastal waters, with refinement ratios $(r_{x,y}^2, r_{x,y}^3, r_{x,y}^4) = (4, 16, 32)$. With $r_{x,y}^1 = 16$, these ratios resulted in a total refinement factor of 2^{15} in each direction, yielding level-5 grids with approximately 5 m resolution in a global-scale computational domain.

Tsunami modeling illuminates the importance of well-balanced methods, described in Sect. 3.2. In the deep ocean, tsunami amplitudes on the order of 1 m, with a wavelength of ≈ 100 km imply that the surface elevation varies by centimeters or less between 10 km grid cells. In the same distance, the seafloor elevation can vary by thousands of meters. Because the steady state arises from the balance of topography gradients on the right-hand side of Eq. (2.4) and depth gradients on the left-hand side of Eq. (2.4), tsunamis are in fact a tiny perturbation to the motionless steady state. At practical grid resolutions, the use of methods that are not well balanced can produce spurious waves much larger than the actual tsunami. Transforming Eq. (2.4) to a nonconservative system for η, u, and v can diminish this problem. However, these nonconservative systems are not appropriate for shock-capturing in the near-shore domain [66].

4.2 Flood Modeling: The 1959 Malpasset Dam Failure

Traditionally, flood prediction has been done without solving governing PDEs, but increasingly, engineers are using codes that solve the shallow water equations (2.4), typically on terrain-fit meshes (see, e.g., [67, 68]). Designing meshes that

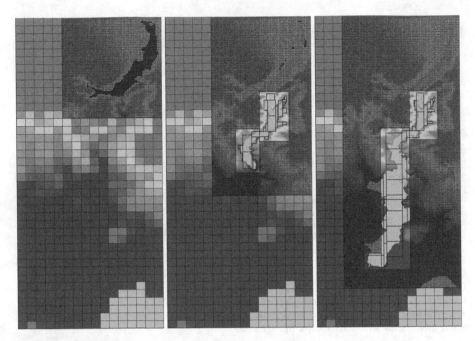

Fig. 9 A GEOCLAW simulation of the Malpasset dam-break flood, described in more detail in [18]. Grids evolve to track the flood as it progresses from the reservoir (*upper right*) toward the Mediterranean Sea (*lower right*). Grid lines are shown only for level-1 and level-2 grids. Individual level-3 and level-4 grids are outlined (color figure online)

conform to the topography of a region can be difficult and time-consuming, and the results obtained are still less than ideal because the optimal resolution evolves in time. Because floods can spread widely through variable terrain, AMR confers an enormous advantage by allowing grids to track a flood along its evolving route. Additionally, as described in [18], well-balanced methods and methods that correctly solve the wet–dry-interface Riemann problem help to accurately resolve the edge of the flood and prevent a spurious numerical sloshing effect that can occur at flow boundaries.

As described in [18], GEOCLAW was used to simulate the flood resulting from the Malpasset Dam failure that occurred on the Côte d'Azur in southern France. This 60-meter-high arch dam failed catastrophically on 2 December 1959, sending a rushing flood into the Reyran River Valley and killing 421 people, mostly in the nearby town of Fréjus [69]. This disaster has served as a benchmark problem for model validation due to the extensive survey work that documented the extent of flooding and timing of the flood arrival.

Results from the GEOCLAW simulation are shown in Figs. 9 and 10. A more detailed description and comparison to field data is provided in [18]. Generally, simulations using the shallow water equations compare surprisingly well to the actual event (see, e.g., [18, 68, 70]). As in the tsunami simulations described above,

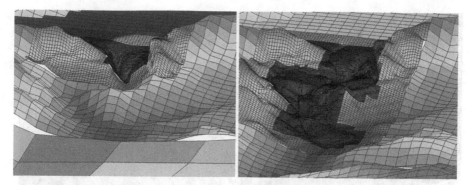

Fig. 10 Close-up oblique view of the Malpasset Dam moments after the initial failure. Level-3 and level-4 grids evolve with the advancing flood waves (color figure online)

a very coarse level-1 grid comprises the domain, which is roughly 5 km by 10 km. The reservoir is resolved on a level-2 grid. Level-3 and level-4 grids then resolve the flood wave as it winds down into the river valley toward the Mediterranean Sea. The refinement ratios were $(r^1_{x,y}, r^2_{x,y}, r^3_{x,y}) = (8,4,4)$, yielding level-4 grids with 3-meter resolution. Figure 10 shows a close-up oblique view near the dam seconds after the dam failure, showing the high-level grids tracking the winding flood path.

4.3 Granular-Fluid Flow Modeling

The depth-averaged hyperbolic model described in [9] was developed for granular-fluid flows, for simulating a variety of phenomena that can be broadly classified as debris flows (volcanic lahars, shallow water-saturated landslides, shallow mudflows, etc.); see, e.g., [27] for an overview of these phenomena. This granular-fluid model has been implemented in GEOCLAW for simulating natural debris flows in mountainous topography, to be described in future publications. GEOCLAW's AMR capability is utilized for these problems in a manner analogous to that described in Sect. 4.2, where grids evolve and follow the flow along its winding route.

In order to rigorously test this granular-fluid model, a 1D numerical code (DIGCLAW) has been developed for validation comparisons with experimental data collected at the U.S. Geological Survey's debris-flow flume in Oregon (Fig. 11). This ≈ 90 m concrete flume is equipped with instrumentation allowing the collection of data during controlled field-scale experiments (see, e.g., [71]). In addition to flow depth, the sensors measure the basal normal stress and pore-fluid pressure as the flow evolves, revealing fundamental feedback processes controlling flow mobility.

There are two principal types of experiments conducted at the debris-flow flume. In one type (e.g., [71]), loose sediment (composed of gravel, sand, and silt) is deposited in a hopper behind a closed steel headgate. The sediment is saturated and flows immediately when the controlled doors are released. In these gate-release

Fig. 11 The USGS debris-flow flume. Instrumentation allows the collection of flow depth, basal normal stress, and pore-fluid pressure during controlled experiments

experiments, the flow is initially far from equilibrium and the unstable mass fails instantaneously. Most numerical models are initialized in a similar way—as a dam-break far from equilibrium. The second type of experiment (e.g., [72]) consists of sediment deposited on the flume bed and slowly saturated in a controlled manner. The mixture can fail spontaneously if the pore-fluid pressure reaches a critical threshold and reduces the shear strength to an unstable level. These mobilization experiments are more representative of natural debris-flow initiation, in which a balanced equilibrium is slightly perturbed, leading to a propagating feedback between grain shearing and changes in pore-fluid pressure, possibly resulting in a dynamic flow [36].

A key motivation in the development of the granular-fluid model [9] is to enable capturing the transition from failure to flow mathematically and numerically. Although a full description of the mathematical model and numerical scheme is beyond the aim of this section, modeling the transition described above provides another example of the importance of well-balanced numerical methods. Much like a tsunami, debris flow motion is often characterized by a near-balanced steady state

$$\mathcal{A}(q)q_x \approx \psi(q,x). \tag{4.19}$$

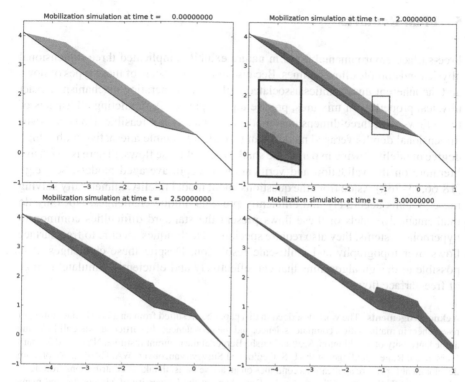

Fig. 12 Preliminary 1D simulations of debris-flow mobilization. *Upper left*: a vertical cross-section of dry debris is shown (*lightly shaded*). As the pore-fluid pressure rises, the shear strength is reduced. Pore pressure is indicated by the *dark shading*, scaled linearly between zero (*no shading*) and lithostatic (entire depth is *shaded*). When the pressure reaches a critical value, the material begins to shear at some location. This can be seen as a small disturbed region in the *upper right panel* near $x = -1.0$ (enlarged in the inset). This instability leads to a rapidly expanding feedback of shearing, grain consolidation, increasing pore pressure, and eventual motion of the entire mass (*lower panels*). In these numerical experiments, the initial fluid pressure rise was dictated numerically prior to motion, after which it evolved spontaneously

However, for a debris flow, large driving forces on the left-hand side of Eq. (4.19) are balanced by high shear resistance and friction on the right-hand side. Initial motion is a small perturbation to this balance that must be accurately captured, particularly in order to correctly resolve the potential transition to rapid mobility or alternatively restabilization and a return to a motionless steady state.

Preliminary use of DIGCLAW to simulate flume experiments indicates that it is capable of capturing these processes. Because the stress terms in the model equations evolve, a sediment mass can remain stable at the beginning of a simulation and potentially fail as the pore pressure and volume fractions coevolve. Alternatively, a flowing mass can stabilize and become motionless. In contrast, models that use a static friction law (constant Coulomb friction) or rate-dependent rheology (e.g., a viscoplastic fluid) must either produce flow initially or remain motionless permanently. Some preliminary numerical results using the 1D DIGCLAW code are shown in Fig. 12.

5 Final Remarks

Free-surface environmental flows in nature exhibit complicated three-dimensional physics and complex flow regimes. Because of the large scale of these types of flows and the inherent uncertainties associated with them (generation mechanisms, exact physical properties of mixtures, precise topography, etc.), predicting all aspects of these flows with three-dimensional models is typically not feasible. However, two-dimensional depth-averaged models can provide a tractable alternative with a high degree of fidelity, owing in part to the shallowness of these flows. (There is extensive literature on the validation and verification of depth-averaged models; see, e.g., [18,66,73–76].) Aside from the questions about model fidelity, numerically solving these systems accurately and efficiently presents unique challenges. Not only do mathematical models of these flows present the standard difficulties common to hyperbolic systems, they also require specialized techniques specific to free-surface flows over topography and multiscale resolution. Despite these challenges, it is possible to design algorithms that can effectively and efficiently simulate a range of free-surface flows.

Acknowledgements The work described in this paper has resulted from an active collaboration of researchers in mathematics, computer science, and the geosciences. In particular, Randall LeVeque at the University of Washington, Seattle; Marsha Berger at the Courant Institute, NYU; and Richard Iverson and Roger Denlinger at the U.S. Geological Survey, Vancouver, WA. GEOCLAW software development for future applications continues on an active basis, thanks to contributions from Kyle Mandli at the University of Texas, Austin; Dave Yuen at the University of Minnesota; and many others.

References

1. J.R. Mehelcic, J.C. Crittenden, M.J. Small, D.R. Shonnard, D.R. Hokanson, Q. Zhang, H. Chen, S.A. Sorby, V.U. James, J.W. Sutherland, J.L. Schnoor, Environ. Sci. Technol. **37**, 5314 (2003)
2. J.J. Stoker, *Water Waves: The Mathematical Theory with Applications* (Interscience Publishers, New York, 1957)
3. G. Whitham, *Linear and Nonlinear Waves* (John Wiley and Sons, New York, 1974)
4. D.T. Resio, J.J. Westerink, Physics Today pp. 33–38 (2008)
5. K.T. Mandli, Finite volume methods for the multilayer shallow water equations with applications to storm surges. Ph.D. thesis, University of Washington, Seattle, WA (2011)
6. R. Abgrall, S. Karni, SIAM Journal on Scientific Computing **31**(3), 1603 (2009)
7. F. Bouchut, V. Zeitlin, Discrete and continuous dynamical systems-series B pp. 739–758 (2010)
8. E.B. Pitman, L. Le, Phil. Trans. R. Soc. A **363**, 1573 (2005)
9. D.L. George, R.M. Iverson, in *The 5th intl. conf. on debris-flow hazards*, ed. by R. Genevois, D. Hamilton, A. Prestininzi (Italian Journal of Engineering, Geology and Environment, Padova, Italy, 2011), pp. 415–424
10. R.J. LeVeque, *Finite Volume Methods For Hyperbolic Problems*. Texts in Applied Mathematics (Cambridge University Press, Cambridge, 2002)
11. G. Dal Maso, P.G. LeFloch, F. Murat, J. Math. Pures Appl. **74**, 483 (1995)

12. L. Gosse, Math. Comp. **71**, 553 (2002)
13. M.J. Castro-Díaz, T. Chacón Rebollo, E.D. Fernández-Nieto, C. Parés., SIAM J. Sci. Comput. **29**, 1093 (2007)
14. R.J. LeVeque, Journal of Scientific Computing **48**, 209 (2010)
15. D.L. George, J. Comput. Phys. **227**(6), 3089 (2008). DOI 10.1016/j.jcp.2007.10.027
16. V.T. Chow, *Open Channel Hydraulics* (McGraw-Hill, 1959)
17. D.L. George, Finite volume methods and adaptive refinement for tsunami propagation and inundation. Ph.D. thesis, University of Washington (2006)
18. D.L. George, Int. J. Numer. Meth. Fluids **66**(8), 939 (2011). DOI 10.1002/fld.2298
19. A.E. Green, P. Naghdi, Journal of Fluid Mech. **78**, 237 (1976)
20. P.K. Stansby, J.G. Zhou, Int. J. Numer. Meth. Fluids **28**, 541 (1998)
21. Z. Li, B. Johns, Int. J. Numer. Meth. Fluids **35**, 299 (2001)
22. J. Sainte-Marie, M. Bristeau, DCDS (B) **10**(4), 733 (2008)
23. Y. Yamazaki, Z. Kowalik, K. Cheung, Int. J. Numer. Meth. Fluids **61**, 473 (2008)
24. E. Audusse, M. Bristeau, B. Perthame, J. Sainte-Marie, Mathematical Modelling and Numerical Analysis (2009)
25. M.J. Castro-Díaz, E.D. Fernández-Nieto, J.M. González-Vida, C. Parés., Journal of Scientific Computing **48**, 16 (2010)
26. T. Takahashi, Annual review of fluid mechanics **13**, 57 (1981)
27. R.M. Iverson, Rev. Geophys. **35**(3), 245 (1997)
28. S.B. Savage, K. Hutter, J. Fluid Mech. **199**, 177 (1989)
29. S.B. Savage, K. Hutter, Acta Mech. **86**, 201 (1991)
30. A.M. Johnson, *Physical Processes in Geology* (W. H. Freeman, 1970)
31. R.M. Iverson, R.P. Denlinger, J. Geophys. Res. **106**(B1), 537 (2001)
32. R.P. Denlinger, R.M. Iverson, J. Geophys. Res. **109**, F01014 (2004). DOI10.1029/2003 JF000085
33. J. Kowalski, Two-phase modeling of debris flows. Ph.D. thesis, ETH Zurich (2008)
34. M. Pelanti, F. Bouchut, A. Mangeney-Castelnau, J.P. Vilotte., in *Hyperbolic Problems: Theory, Numerics, Applications.*, ed. by S. Benzoni-Gavage, D. Serre (Springer, 2008). Proc. 11th Intl. Conf. on Hyperbolic Problems, Lyon France, July 2006.
35. M. Pelanti, F. Bouchut, A. Mangeney, Mathematical Modelling and Numerical Analysis **42**(5), 851 (2008)
36. R.M. Iverson, J. Geophys. Res. **110**, F02015 (2005)
37. T.W. Lambe, R.V. Whitman, *Soil Mechanics* (John Wiley and Sons, 1969)
38. M.J. Castro-Díaz, E.D. Fernández-Nieto, A.M. Ferreiro, Comput. Fluids **37**, 299 (2008)
39. J. Murillo, J. Burguete, P. Brufau, P. García-Navarro, Int. J. Numer. Meth. Fluids **49**, 267 (2005). DOI 10.1002/fld.992
40. T. Morales de Luna, M.J. Castro Díaz, C. Parés, E.D.F. Nieto, Communications in Computational Physics **6**, 848 (2009)
41. P. Lynett, P.L.F. Liu, Proceedings of the Royal Society of London A **458**, 2885 (2002)
42. M.J. Berger, D.L. George, R.J. LeVeque, K. Mandli, Advances in Water Resources p. in press (2011). DOI 10.1016/j.advwatres.2011.02.016
43. E.F. Toro, *Riemann Solvers and Numerical Methods for Fluid Dynamics* (Springer–Verlag, Berlin, 1997)
44. R.J. LeVeque, J. Comput. Phys. **131**, 327 (1997)
45. V.V. Titov, C.E. Synolakis, Jounal of Waterways, Ports, Coastal and Ocean Engineering **124**(4), 157 (1998)
46. J.M. Gallardo, C. Pares, M. Castro, J. Comput. Phys. **227**, 574 (2007)
47. F. Marche, P. Bonneton, P. Fabrie, N. Seguin, Int. J. Numer. Meth. Fluids **53**, 867 (2007)
48. P.L. Roe, J. Comput. Phys. **43**, 357 (1981)
49. A. Harten, P.D. Lax, B. van Leer, SIAM Review **25**, 235 (1983)
50. R.J. LeVeque, J. Comput. Phys. **146**, 346 (1998)
51. J.F. Colombeau, *Multiplication of Distributions: A tool in mathematics, numerical engineering and theoretical physics* (Springer Verlag, Berlin, 1992)

52. T. Gallouët, J.M. Hérard, N. Seguin, Comput. Fluids **32**, 479 (2003)
53. M.J. Castro, P.G. LeFloch, M.L. Munoz, C. Parés., J. Comput. Phys. **227**, 8107 (2008)
54. J.M. Greenberg, A.Y. LeRoux, SIAM J. Numer. Anal. **33**, 1 (1996)
55. P. García-Navarro, M.E. Vázquez-Céndon, Comput. Fluids **29**, 17 (2000)
56. L. Gosse, Math. Mod. Meth. Appl. Sci. **11**, 339 (2001)
57. F. Bouchut, *Nonlinear Stability of Finite Volume Methods for Hyperbolic Conservation Laws and Well-Balanced Schemes for Sources* (Birkhäuser Verlag, 2004)
58. E.F. Toro, *Shock Capturing Methods for Free Surface Shallow Flows* (John Wiley and Sons, Chichester,United Kingdom, 2001)
59. E. Audusse, M.O. Bristeau, J. Comput. Phys. **206**, 311 (2005)
60. A. Kurganov, G. Petrova, Commun. Math. Sci. **5**(5), 133 (2007)
61. E. Audusse, F. Bouchut, M.O. Bristeau, R. Klein, B. Perthame, SIAM J. Sci. Comput. **25**, 2050 (2004)
62. R. Bernetti, V. Titarev, E. Toro, J. Comput. Phys. **227**, 3212 (2008)
63. M.J. Berger, J. Oliger, J. Comput. Phys. **53**, 484 (1984)
64. M.J. Berger, P. Colella, J. Comput. Phys. **82**, 64 (1989)
65. M.J. Berger, R.J. LeVeque, SIAM J. Numer. Anal. **35**, 346 (1998)
66. R.J. LeVeque, D.L. George, M.J. Berger, Acta Numerica **20**, 211 (2011). DOI10.1017/S09 62492911000043
67. J.M. Hervouet, A. Petitjean, J. Hydraul. Res. **37**, 777 (1999).
68. A. Valiani, V. Caleffi, A. Zanni, J. Hydraul. Eng. **128**(5), 460 (2002)
69. USBOR, Arch dam failures: Malpasset and St. Francis. Tech. rep., Unites States Bureau of Reclamation (unknown)
70. D. Pianese, L. Barbiero, J. Hydraul. Eng. **128**(5), 941 (2004)
71. R. Iverson, M. Logan, R. LaHusen, M. Berti, J. Geophys. Res. **115**, 1 (2010)
72. R.M. Iverson, M.E. Reid, N.R. Iverson, R.G. LaHusen, M. Logan, J.E. Mann, D.L. Brien, Science **290**(5491), 513 (2000)
73. A. Mangeney, P. Heinrich, R. Roche, Pure appl. Geophys. **157**, 1081 (2000)
74. R.P. Denlinger, M. Iverson, J. Geophys. Res. **106**, 553 (2001)
75. M. Briggs, C. Synolakis, G. Harkins, Proc. of Waves—Physical and Numerical Modeling (1994)
76. V.V. Titov, C.E. Synolakis, Tsunami 93 Proc. IUGG/IOC Intl Tsunami Symp. pp. 627–636 (1993)

Real-Time Forecasting and Visualization of Hurricane Waves and Storm Surge Using SWAN+ADCIRC and FigureGen

J.C. Dietrich, C.N. Dawson, J.M. Proft, M.T. Howard, G. Wells, J.G. Fleming,
R.A. Luettich Jr., J.J. Westerink, Z. Cobell, M. Vitse, H. Lander,
B.O. Blanton, C.M. Szpilka, and J.H. Atkinson

Abstract Storm surge due to hurricanes and tropical storms can result in significant loss of life, property damage, and long-term damage to coastal ecosystems and landscapes. Computer modeling of storm surge is useful for two primary purposes: forecasting of storm impacts for response planning, particularly the evacuation of vulnerable coastal populations; and hindcasting of storms for determining risk, development of mitigation strategies, coastal restoration, and sustainability. Model

J.C. Dietrich (✉) • C.N. Dawson • J.M. Proft • M. Vitse
Institute for Computational Engineering and Sciences, University of Texas at Austin,
201 East 24th Street, C0200, Austin, TX 78712, USA
e-mail: dietrich@ices.utexas.edu

M.T. Howard • G. Wells
Center for Space Research, University of Texas at Austin, 201 East 24th Street,
C0200, Austin, TX 78712, USA

J.G. Fleming
Seahorse Coastal Consulting, Morehead City, NC, USA

R.A. Luettich Jr.
Institute of Marine Sciences, University of North Carolina at Chapel Hill,
Chapel Hill, NC, USA

J.J. Westerink
Department of Civil and Environmental Engineering and Earth Sciences,
University of Notre Dame, South Bend, IN 46556, USA

Z. Cobell • J.H. Atkinson
Arcadis Inc., Denver, CL, USA

H. Lander • B.O. Blanton
Renaissance Computing Institute, University of North Carolina at Chapel Hill,
Chapel Hill, NC, USA

C.M. Szpilka
School of Civil Engineering and Environmental Science, University of Oklahoma,
Norman, OK 73019, USA

C. Dawson and M. Gerritsen (eds.), *Computational Challenges in the Geosciences*, The IMA   49
Volumes in Mathematics and its Applications 156, DOI 10.1007/978-1-4614-7434-0_3,
© Springer Science+Business Media New York 2013

results must be communicated quickly and effectively, to provide context about the magnitudes and locations of the maximum waves and surges in time for meaningful actions to be taken in the impact region before a storm strikes.

In this paper, we present an overview of the SWAN+ADCIRC modeling system for coastal waves and circulation. We also describe FigureGen, a graphics program adapted to visualize hurricane waves and storm surge as computed by these models. The system was applied recently to forecast Hurricane Isaac (2012) as it made landfall in southern Louisiana. Model results are shown to be an accurate warning of the impacts of waves and circulation along the northern Gulf coastline, especially when communicated to emergency managers as geo-referenced images.

Keywords Hurricane waves • Storm surge • Hurricane Isaac (2012) • ASGS • SWAN • ADCIRC • FigureGen

1 Introduction

Storm surge is primarily a competition between wind and wave forcing and frictional resistance. As hurricanes approach the coast, water is driven inland and can cause significant flooding, loss of life, and damage to property and coastal ecosystems. Predicting and understanding the magnitude and geographic extent of surge is critical to emergency managers in the event of an impending landfall, and to longer-term efforts to protect and sustain coastal environments. Computer models of storm surge are central to these efforts.

The modeling of hurricane waves and storm surge has advanced significantly in the last decade, motivated by the active seasons of 2004 (Charley, Frances, Ivan), 2005 (Katrina, Rita), 2008 (Gustav, Ike), and 2012 (Isaac), all of which caused significant flooding along the U.S. Gulf of Mexico coastline from Texas to Florida. To simulate the waves and surge caused by these storms, model advancements have included the improved representation of wind stress forcing, incorporation of dynamic inflows from major waterways, such as the Atchafalaya and Mississippi Rivers, and better parameterizations of input and boundary conditions. Tighter coupling between the wave and circulation models has also improved the simulation of these processes, notably within the coupling of the Simulating WAves Nearshore (SWAN) and ADvanced CIRCulation (ADCIRC) models [14, 15]. The resulting SWAN+ADCIRC model simulates hurricane waves and storm surge from deep water to the nearshore in a manner shown to be both accurate and efficient [17].

The SWAN+ADCIRC model benefits from improvements in the computational meshes that describe the Gulf Coast. Complex coastal regions contain channels, levees, raised roads, and other internal barriers that must be included in a description of the model domain, since they either enhance or impede inland flow. The computational domain is discretized using triangular finite elements, to better represent complex coastal features, barrier islands, and internal barriers, and to allow for gradation of the mesh that increases feature detail in moving from the

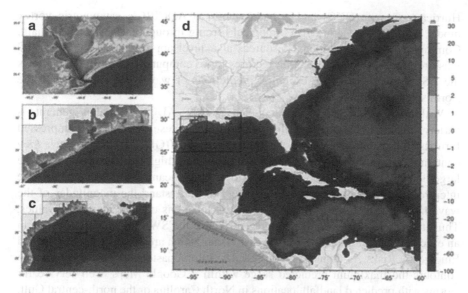

Fig. 1 Details of the bathymetry/topography (m) in the TX2008r35h mesh, with panels at successive zoom levels indicated by *black rectangles*. The depth/elevation contour range is identical in all panels. The panels depict the geographic regions of: (**a**) Galveston Bay, (**b**) the north Texas coast, (**c**) the Louisiana-Texas continental shelf, and (**d**) the entire computational domain

deeper ocean, onto the continental shelf, into estuaries and marshes, and over low-lying coastal floodplains. These meshes have evolved and been validated for storm applications in southeastern Louisiana [8, 15, 40] and Texas [10, 22]. Recent meshes contain millions of triangular elements and feature mesh spacings that range from 4–6 km in the deeper Gulf, 500–1,000 m on the continental shelf, 200 m within the coastal floodplains, and downward to 20 m within the fine-scale natural and manmade channels and levees. In Fig. 1, the multiple scales are illustrated within a high-resolution mesh of the Texas coastline. These high-resolution meshes provide accurate representations of the waves and circulation characteristics within the coastal environment, but their effective use requires numerical algorithms that are efficient in parallel computing environments [17, 36].

The coupled SWAN+ADCIRC models are applied in real time to generate forecasts of hurricane waves and storm surge via their implementation in the ADCIRC Surge Guidance System (ASGS) [33, 34]. The ASGS was created initially in 2006 to provide guidance to the U.S. Army Corps of Engineers following the construction of gates along the rainwater outfall canals on the north side of New Orleans after Hurricane Katrina; the operation of the new gates depended on the timing and severity of wind speed and storm surge within Lake Pontchartrain as storms approached [18]. The ASGS has been adapted continuously to allow for portability to disparate computing environments, geographical relocation to different computing sites, and flexible postprocessing. The ASGS parses storm parameters from official forecast/advisories issued at 6-h intervals from the National

Hurricane Center (NHC) and provides them to ADCIRC's asymmetric vortex model [20, 27] to generate meteorological forcing throughout the model domain. Commitment of computational resources at a level appropriate to the ADCIRC mesh resolution (higher resolution requires more computational resources) allows the ASGS to provide high-resolution predictions of waves and surge during the storm's approach to the coast.

Within the ASGS, the tightly coupled SWAN+ADCIRC models generate a variety of output files to describe their simulated results. Even in compressed formats, these output files can be as large as 5–10 GB, especially when they contain days of information for a simulation using a high-resolution mesh. These large output files must be postprocessed to provide meaningful results that can be interpreted quickly by emergency managers to understand the magnitudes of the wave heights and storm surge, as well as the locations of their maximum values. Thus it is essential to visualize and geo-reference the SWAN+ADCIRC results in an efficient manner. Examples of forecast visualization are provided by the Coastal Emergency Risks Assessment group (CERA, http://coastalemergency.org/), which contours the maximum predicted surge within a Google Maps Web service for storms with predicted landfall locations in North Carolina or the north-central Gulf. These maps are interactive, but their content is fixed, and thus the user cannot include additional layers of visualization from other sources.

Another visualization tool is FigureGen, which has been developed for illustration of the input and output files that describe SWAN+ADCIRC simulations (http://www.caseydietrich.com/figuregen/). FigureGen creates illustrations of unstructured meshes, bathymetry/topography, input parameterizations such as Manning's n values, and computed quantities such as water levels and significant wave heights. It can be implemented in a parallel computing environment, so that the cores work together to illustrate multiple time frames from an output file, and so that the output files can be processed immediately as they are written by the wave and circulation models. FigureGen creates publication-quality images in raster graphics formats including TIFF, JPG, and PNG, and it also geo-references images for use with software packages such as Esri's ArcGIS and Google Earth. The open-source FORTRAN code is available for all applications of SWAN+ADCIRC, and the simplicity of its execution and parameter file format mean that it is easily scriptable. As a result, the generation of visualizations with FigureGen offered a simple addition to the flexible postprocessing facility of the ASGS.

An example of the forecasting application occurred during Hurricane Isaac (2012), which crossed the Gulf of Mexico during late August before making landfall in southeastern Louisiana as a Category 1 storm. The storm's projected track was uncertain, especially as it moved through the Caribbean Sea. The predicted landfall location shifted along the Florida coastline, appeared to stabilize on the Florida panhandle near Pensacola, and then moved to southeastern Louisiana less than 48 h before the eventual landfall. Because of this track uncertainty, there was intense concern about the storm from forecasters throughout the Gulf. In particular, the lead authors of the present study were tasked with providing forecast guidance to the Texas State Operations Center (SOC), and these results were shared with the

National Weather Service Southern Region Headquarters in Fort Worth, Texas, as well as local NWS Forecast Offices in Tallahassee and Miami, Florida. To make the best use of available computing resources, these forecasts were performed on a coarsely resolved mesh with coverage of the entire Gulf. As will be shown below, these forecast results provided an efficient approximation of the waves and surge along the Gulf coastline, especially when communicated with geo-referenced images created by FigureGen.

The following sections describe components of the ASGS and the results of operational support using ASGS during Isaac. Section 2 summarizes the SWAN and ADCIRC models for waves and circulation, including references to detailed descriptions of their numerics, input, and boundary conditions; the unstructured meshes are also described. Section 3 describes the FigureGen tool, including its input file types, the software tools it uses to generate its images, its implementation in parallel computing environments, and the options for geo-referenced output products. Finally, Sect. 4 describes the implementation of these models for real-time forecasting of Isaac, with an emphasis on large-scale validation and examples of the output products that were shared with weather forecasters and emergency managers during the event.

2 Models of Waves and Circulation in the Gulf of Mexico

2.1 Tight Coupling of SWAN+ADCIRC

In large-scale applications, it can become inefficient to resolve the phases of individual waves, and thus the SWAN and other phase-averaged wave models consider the evolution of action density $N(t, \lambda, \varphi, \theta, \sigma)$ in time t, geographic space (λ, φ), and spectral space with directions θ and frequencies σ [5]. The action density N can be integrated to determine properties of the wave environment, such as significant heights and mean periods. Wave energy is generated, propagated, and dissipated via source terms that represent wave growth by wind; action lost due to whitecapping, surf breaking, and bottom friction; and action exchanged between spectral components due to nonlinear effects in deep and shallow water. The source term parameterizations used herein are identical to those in recent studies [15, 17].

SWAN employs the finite difference method, with a third-order upwind scheme for the advection terms in geographic space [35] and a diffusive correction for the "garden-sprinkler" effect [4]. A Gauss–Seidel iterative technique is employed to update the action densities in geographic space, by ordering the vertices and then sweeping through them in opposite directions. This solution method is implicit and thus unconditionally stable. Detailed descriptions of the SWAN solution method are available [5]. This method was extended recently to utilize unstructured meshes with triangular elements [41], so mesh resolution can be improved in regions with large gradients in input parameters (e.g., bathymetry) or the computed solution.

The ADCIRC model has been validated using tide gauges and field measurements for several hurricanes in southern Louisiana [8, 13, 40], and it has been used extensively by the U.S. Army Corps of Engineers (USACE), the Federal Emergency Management Agency (FEMA), and local agencies to design flood control systems and to evaluate hurricane flooding risk.

ADCIRC computes water levels ζ and depth-averaged currents (U, V) via solution of modified forms of the shallow-water equations (SWE) [9, 23, 25, 40]. The model applies the continuous-Galerkin, finite-element method with linear C_0 triangular elements to discretize and solve the SWE on unstructured meshes. Water levels ζ are determined from the Generalized Wave Continuity Equation (GWCE), which is a combined and differentiated form of the continuity and momentum equations, and which is discretized over three time intervals, so that the solution for the future water level requires knowledge of the present and past water levels. Current velocities (U, V) are determined from the vertically integrated momentum equations, which are discretized explicitly for all terms except the Coriolis force, which uses an average of the present and future velocities [36].

ADCIRC utilizes boundary conditions and input parameterizations to simulate effectively the circulation in coastal regions. The specific details of these parameters are contained in other publications, and thus they are only referenced herein. Large rivers with significant impacts on the coastal circulation, such as the Mississippi and Atchafalaya Rivers in Louisiana, are forced with an inflow boundary condition that prevents surge and tidal waves from reflecting back into the computational domain [26, 40]. Tidal constituents are forced along the open-ocean boundary in the Atlantic Ocean, and tidal potential functions are forced within the model domain [24, 28]. Bottom friction is parameterized with Manning's n values that vary spatially based on representative land-cover data [1, 2, 7, 8]. Wind stresses are computed using a quadratic drag law [19] with adjustments based on storm sector [15, 31, 32]. Wind stresses are corrected in overland regions depending on upwind roughness and the presence of tree canopies [40]. Overland regions are allowed to flood or dry as the storm surge inundates or recedes, respectively [12].

The SWAN and ADCIRC models are coupled tightly so that they run as the same executable and on the same unstructured meshes [14, 17]. Coupling information is passed through local memory/cache without the need for interpolation between heterogeneous meshes. ADCIRC passes wind velocities, water levels, current velocities, and friction roughness lengths to SWAN, while SWAN passes wave radiation stress gradients to ADCIRC. The coupling interval is taken to be the same as the SWAN time step, which is 20 min for the current simulations.

In a high-performance computing environment, the domain is decomposed into local subdomains for the computational cores [21]. Each core runs the coupled models on a local submesh, with an overlapping layer of elements that allows for intramodel communication between neighboring submeshes. However, the intermodel communication between SWAN and ADCIRC is still intracore via local memory/cache, because there is no need to interpolate values over the network. Thus the tight coupling is highly efficient. SWAN+ADCIRC maintains the excellent

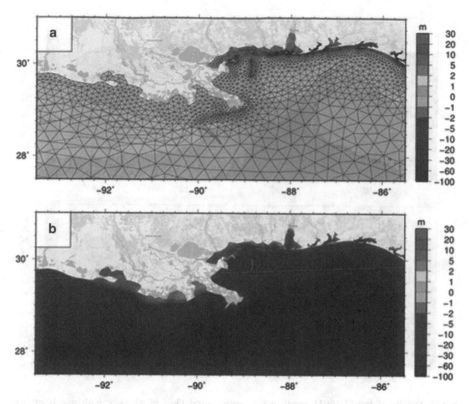

Fig. 2 Details of the EC95d mesh in the north-central Gulf of Mexico, with panels of (**a**) unstructured mesh and (**b**) bathymetry and topography relative to NAVD88 (2004.65). Note that the full computational mesh extends through the Gulf of Mexico and Caribbean Sea and to 60° W longitude in the Atlantic Ocean

scalability of its component models, and provides high-resolution simulations of hurricane waves and storm surge in less than 10 min per day of simulation [17, 36].

2.2 Unstructured Meshes

The SWAN+ADCIRC models have been validated on unstructured meshes within the Gulf of Mexico region, with varying levels of resolution, for applications ranging from tidal databases to hurricane forecasts. Examples of coarsely and finely resolved meshes are shown in Figs. 2 and 3, respectively.

The Eastcoast 1995 (EC95d) mesh was developed for the generation of a tidal database in the Western North Atlantic Ocean, based on earlier studies for mesh generation for tidal computations [38, 39]. The EC95d mesh contains 31,435 vertices and 58,369 triangular elements, and it includes coverage of the entire western North

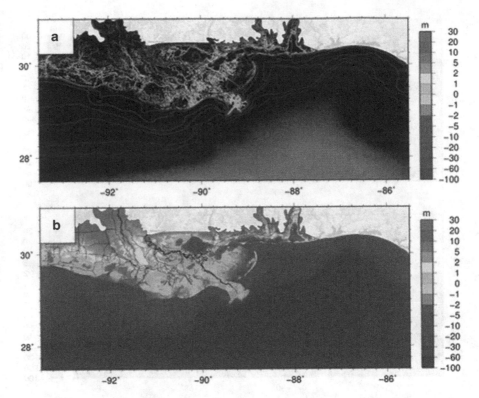

Fig. 3 Details of the SL16v31 mesh in the north-central Gulf of Mexico, with panels of (**a**) unstructured mesh and (**b**) bathymetry and topography relative to NAVD88 (2004.65). Note that the full computational mesh extends through the Gulf of Mexico and Caribbean Sea and to 60°W longitude in the Atlantic Ocean

Atlantic Ocean, Caribbean Sea, and Gulf of Mexico. Its coarse spatial resolution ranges upward to element sizes of 100 km, and it does not include any coastal floodplains (Fig. 2). In the northern Gulf, the mesh spacings range downward to 5–10 km on the continental shelf and 1–2 km along the coastlines of Mississippi and Alabama. Thus, although the EC95d mesh contains sufficient levels of mesh resolution to propagate the tides into the Gulf, it was not designed for application to nearshore hurricane waves and surge. However, its coarse mesh resolution allows faster forecasts when storms are located far from the coastline of interest.

This mesh was improved and refined, and its boundary was extended inland to allow for coastal inundation, resulting in a series of meshes with progressively greater spatial resolution [8, 28, 40]. SWAN+ADCIRC was validated recently on the Southern Louisiana (SL16v31) mesh for high-resolution hindcasts of the recent storms to impact the northern Gulf, including Katrina and Rita (2005) and Gustav and Ike (2008) [15, 17]. The SL16v31 mesh contains 5,035,113 vertices and 9,945,623 elements, and thus it is roughly 160 times the size of the EC95d mesh. It contains mesh spacings of 20–25 km in the Caribbean Sea and Atlantic

Ocean, 4–6 km in the Gulf of Mexico, 500–1,000 m on the continental shelf, and 200 m or smaller in the coastal floodplains of southern Louisiana, Mississippi, and Alabama (Fig. 3). The mesh spacings range downward to 20 m in natural and manmade channels such as the distributaries of the Mississippi River Delta. When applied on this mesh, SWAN+ADCIRC has been shown to provide highly accurate hurricane simulations throughout the northern Gulf, and it has also been shown to be highly efficient, provided that enough computational cores are utilized to maintain a problem size of less than 10,000 mesh vertices per 1 MB of share cache [17]. Thus, the SL16v31 mesh can be employed with SWAN+ADCIRC in a real-time production framework if it is allotted an appropriate commitment of computational resources.

3 FigureGen

The tightly coupled SWAN+ADCIRC models generate a variety of output files in formats including ASCII text, binary, and the Network Common Data Form (NetCDF, http://www.unidata.ucar.edu/software/netcdf/). However, even in compressed formats, output files can be 10 GB or larger, especially when they contain information covering simulated days of model computations using high-resolution meshes. These large output files must be postprocessed to extract meaningful results targeted for emergency managers to understand the threats posed by the magnitudes of the wave heights and storm surge, as well as the locations of their maximum values. Thus it becomes essential to visualize the SWAN+ADCIRC results. This visualization must geo-reference accurately the results within the coastal environment, to provide context for the end user. And this visualization must be efficient, to deliver illustrations in a timely manner. For these reasons, the FigureGen visualization tool was developed to provide high-quality illustrations of SWAN+ADCIRC results.

FigureGen is a FORTRAN program that acts as an interface between the SWAN+ADCIRC simulation files and the resulting illustrations (http://www.caseydietrich.com/figuregen/). The unstructured mesh can be illustrated in terms of its triangular elements and/or contours of mesh spacings or bathymetry, as in Figs. 2 and 3. Vertex-based input parameterizations, such as Manning's n values and directional wind reduction factors, can also be contoured spatially. For parallel computing applications, the domain decomposition can be visualized to show the local submeshes. Background images can be underlaid to show satellite imagery or previous simulation results.

FigureGen also visualizes the SWAN+ADCIRC output files in ASCII and NetCDF formats. ADCIRC produces files with global data sets of water levels, current velocities, atmospheric pressures, wind velocities, and wave radiation stress gradients [36]. SWAN produces files with global data sets of significant wave heights, peak and mean wave periods, and mean wave directions [14]. Scalar and vector data can be plotted with filled and/or linear contours, and vector data can also be overlaid on a regular grid. FigureGen can also visualize the locations of conservative tracers, such as the simulated oil transport following the destruction of the Deepwater Horizon platform in the Gulf [16].

To create the illustrations, FigureGen relies on the Generic Mapping Tools (GMT, http://gmt.soest.hawaii.edu/). GMT is an open-source collection of Unix-based command-line tools for manipulating and visualizing geographic data sets [37]. The tools have undergone more than 20 years of continuous development with support from the National Science Foundation. The GMT suite processes data on structured and unstructured meshes; performs operations such as filtering, trend fitting, and coordinate projection; and produces illustrations with filled and linear contours, vectors, etc., in two or three dimensions.

GMT requires input data in specific gridded formats, and thus FigureGen converts the ADCIRC input and output files before passing them to GMT. FigureGen develops files containing the scalar or vector quantities to be plotted at the locations of the mesh vertices, the vertex connectivity information, the color palettes, etc. Then these files are passed to the GMT command-line tools. It is noted that the FigureGen implementation does not alter the source code of GMT, but rather communicates with the GMT tools via external files, and thus it can be extended to future GMT releases as they become available. At the same time, FigureGen does not require any knowledge of GMT by its user, because the interaction is handled automatically.

In a parallel computing environment, FigureGen uses the Message Passing Interface (MPI) to divide the work among the computational cores. The first core coordinates the overall effort by assigning new images to be produced by the other cores. The visualization cores read the SWAN+ADCIRC files, convert the data into appropriate formats, and then call the GMT tools to generate contours, vectors, etc. In this way, an output file containing numerous discrete time frames can be visualized in parallel, with the work shared by the computational cores. Workflow acceleration is limited only by the number of available cores. FigureGen can also be coupled loosely with SWAN+ADCIRC, by running concurrently and querying the output files for the next time step. As the data is written by the coupled models, it is read and visualized by FigureGen. In this way, the illustrations are ready for the user as the simulation reaches the final time step.

FigureGen creates illustrations in raster image formats such as TIFF, JPG, GIF, and PNG. It can also geo-reference the images for use in other software packages such as Esri's ArcGIS and Google Earth. For example, to output files for use with Google Earth, the image is simplified by removing the labeling around the data frame and moving the contour and vector scales into separate image files. A text file is written in the Keyhole Markup Language (KML), and then everything is compressed into a zipped format (KMZ). These KMZ files allow emergency managers to overlay the SWAN+ADCIRC visualizations on satellite and aerial imagery and other geo-referenced data sets prepared for use in Google Earth. FigureGen can also create vector-based graphics such as encapsulated postscript (EPS) and postscript (PS). These formats are ideal for portable document format (PDF) publications.

4 Hurricane Season 2012

4.1 ASGS Forecasting

Forecast modeling of hurricane waves and storm surge requires a real-time system that is fully automated and resilient, especially within a shared high-performance computing environment. The system must monitor and detect when new data describing meteorological and riverine forcings become available, and download and convert them into formats appropriate for input to the wave and circulation models. It must preprocess the input files and decompose the domain into localized problems for the computational cores. It must submit and monitor each simulation within the batch queue on the computing resource. And it must detect when each simulation finishes, so that the model results can be visualized and shared. For all of these processes, the system should detect and work around errors when possible. In addition, the system should be extensible to a variety of computing environments, to provide redundancy in model forecasts.

For these reasons, the ASGS was developed to automate the use of SWAN+ADCIRC in real-time forecasting environments [18,33,34]. The ASGS has been employed for hurricane applications including Irene (2012) along the North Carolina coastline [3], as well as forecasting during the oil spill resulting from the destruction of the Deepwater Horizon drilling platform in the Gulf [16]. The ASGS accepts meteorological forcing in several formats. Under normal conditions, the system uses gridded wind and pressure files such as the model output from the NOAA National Centers for Environmental Prediction's (NCEP) North American Mesoscale (NAM) model. When a hurricane threatens a coastline, the system downloads the forecast advisories from the NHC, and the storm parameters are used as inputs to generate wind fields using an asymmetric vortex model [20,27]. Riverine influxes are downloaded from the NOAA National Severe Storms Laboratory (NSSL) and used as boundary conditions when necessary. Tidal input and boundary conditions are also developed automatically. The ASGS suite of Perl scripts operates on the front end of a computing cluster, downloads and preprocesses simulation files, monitors the progress of simulations as they run, and produces visualizations of the model results.

The ASGS produced results used by several teams during the 2012 hurricane season. Forecasters from the University of North Carolina (UNC) implemented the system on high-performance computing clusters at the Renaissance Computing Institute (RENCI, http://www.renci.org/), the U.S. Army Engineer Research and Development Center (ERDC, http://www.erdc.hpc.mil/), and other locations. The UNC team focuses primarily on storms with projected impacts along the Carolina coastline. As the lead developers of the ASGS, the UNC team also supports other forecasters using the system, including those at Louisiana State University (LSU), where the focus is primarily on storms with projected impacts along the north-central Gulf coastline. Both teams use the system on high-resolution meshes representing the barrier islands, bays and estuaries, and coastal floodplains in their respective regions of interest. Among other output products, the UNC and LSU teams provide forecast guidance via the CERA (http://coastalemergency.org/), which incorporates the model results within a Google Maps web service.

At the University of Texas at Austin (UT), the lead authors of the present study were tasked with providing forecast guidance to the Texas SOC for storms with projected impacts along the Texas coastline. This forecast modeling is performed typically on a high-resolution mesh (Fig. 1) representing the broad continental shelf, barrier islands, and coastal floodplains ranging from Port Isabel to Port Arthur, Texas; SWAN+ADCIRC has been validated extensively on this mesh for waves and surge during Ike (2008) [10, 22]. The UT team implemented the ASGS at the Texas Advanced Computing Center (TACC http://www.tacc.utexas.edu/). Because of the needs of the UT team and its local partners, the ASGS results for Texas were not posted publicly online, but rather were shared as raster and geo-referenced images. An automated notification system supported the rapid dissemination of images to support decision-makers preparing for storm impacts. In particular, the KMZ files from FigureGen allowed emergency managers to work with the computed results within Google Earth applications, where they can be visualized and overlaid with other data in similar formats. Examples of these output products are shown in subsequent sections.

4.2   Hurricane Isaac

Isaac (2012) differed from other recent storms that made landfall in southeastern Louisiana, such as Katrina (2005) and Gustav (2008), in that it was relatively weaker and slower moving. Isaac passed over Hispaniola and Cuba as a tropical storm, and it was expected to strengthen to a hurricane shortly after it moved into the Gulf. However, although its central pressure reached a minimum of 968 mbar (typical of a Category 2 hurricane on the Saffir–Simpson scale), its core did not become well organized until just before it made landfall, and thus it approached the Louisiana coastline as a Category 1 hurricane. Isaac made initial landfall at 2012/08/28/2345 UTC near the western tip of the Mississippi River Delta, moved offshore near Barataria Bay, and then made a second landfall at 2012/08/29/0700 UTC near Port Fourchon, Louisiana [6, 30]. Slow storm movement caused significant amounts of rainfall, including reports of 20 in within the New Orleans area. And the storm's counterclockwise rotation pushed surge along the Louisiana–Mississippi continental shelf, including reports of 3.4 m of surge measured by a tide gauge near Shell Beach on Lake Borgne east of New Orleans [29]. This storm surge threatened the protection system around New Orleans and caused extensive flooding in communities, such as Braithwaite, Louisiana, exposed outside of the protective coastal infrastructure.

The storm's projected track was uncertain, especially as the storm moved westward through the Caribbean Sea. The predicted landfall location varied along the Florida coastline and then appeared to stabilize on the Florida panhandle near

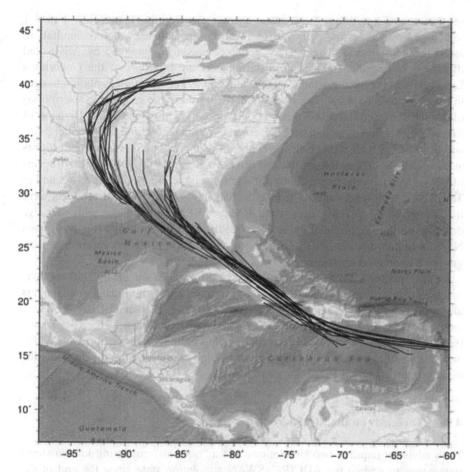

Fig. 4 Consensus 84-h forecast tracks for Isaac (2012) issued by the National Hurricane Center for advisories 1–38. The predicted landfall location was the Florida panhandle for advisories 13–20, but it shifted to Louisiana by advisory 24 (shown in *red*), which was issued 2012/08/27/0300 UTC, or less then 2 days before the initial landfall. Colors refer to figure as printed online

Pensacola, notably during NHC advisories 13–20 (issued over 2 days, August 24–26). As shown in Fig. 4, the projected landfall location shifted westward during August 26 and settled on southeastern Louisiana by NHC advisory 24 (which was issued 2012/08/27/0300 UTC). Because of this track uncertainty, there was intense concern from storm forecasters throughout the Gulf, including the SWAN+ADCIRC modeling teams using the ASGS. The LSU/UNC teams provided guidance based on a high-resolution mesh of southern Louisiana. The lead authors of the present study provided guidance based on the coarsely resolved EC95d mesh.

In the Texas SOC in Austin, an activation of the Governor's Emergency Management Council for Isaac was supported by the UT team. Although Isaac represented a threat primarily to coastal areas in Louisiana and Mississippi, teams of first responders from Texas state agencies, including Texas Task Force 1 search-and-rescue teams, deployed to locations in Louisiana before landfall to assist local responders. The forecast guidance products generated by ASGS and illustrated by FigureGen were used during periodic briefings of the Governor's Emergency Management Council to update agency representatives about changes in the magnitude of impacts to areas offshore and along the coastline. Descriptions of the impacts were recorded in situation reports issued by the Texas Division of Emergency Management to the media, emergency managers, and public officials. In addition, the UT team worked closely with National Weather Service meteorologists at the NWS Southern Region Headquarters in Fort Worth and in local forecast offices in Miami with responsibility for warnings covering the Florida Keys and in Tallahassee with responsibility for Apalachee Bay. ASGS forecast products were transmitted to the NWS offices using KMZ and PNG files along with narrative descriptions that highlighted the results. Feedback received during the event from NWS indicated that the ASGS results were very useful in the preparation of their guidance to emergency managers.

In the sections that follow, the forecast performance on the EC95d mesh is assessed using comparisons with measured time series of significant wave heights and water levels. Then hindcasts are performed on the EC95d and SL16v31 meshes using wind fields generated from the best-track analysis from NHC. It is shown that the EC95d mesh provides an economical approximation of the hurricane waves and storm surge, but higher levels of mesh resolution are required to determine the details of threats posed to specific coastal areas.

4.2.1 Forecasts on the EC95d Mesh

Each advisory requires two simulations: (1) a "nowcast" run using known storm parameters to update the ADCIRC+SWAN simulation state since the end of the last nowcast; and (2) a "forecast" run combining the freshly updated nowcast state and the parameters from the latest official forecast advisory to predict the waves and storm surge 5 days into the future. To accomplish this, the model domain must be decomposed into local subdomains for the computational cores, each of which uses the ADCIRC asymmetric vortex model and the storm parameters provided by the ASGS to generate wind fields in parallel and in real time. When the forecast is finished, the results must be postprocessed and visualized. This process is completed quickly when the ASGS is employed on the EC95d mesh as a result of its modest size. The EC95d mesh resolution requires a relatively small total of 120 cores, and thus these computational jobs may be submitted to the development queue on the TACC Lonestar system, where they will execute immediately, without queue-related delays. The Isaac advisories required an average duration of 26.3 min, of which only 50 s was required by FigureGen to create the Google KMZ visualization products. In summary, the forecast guidance was available within 30 min after each advisory was posted by the NHC when running the EC95d mesh via the ASGS.

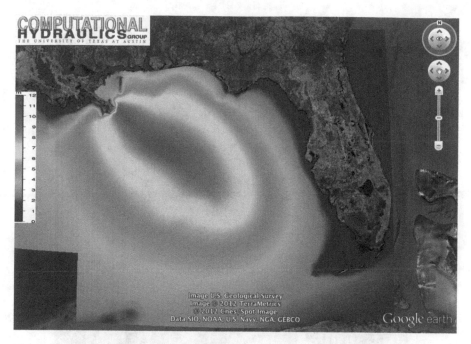

Fig. 5 FigureGen visualization within Google Earth of maximum SWAN-computed significant wave heights on the EC95d mesh for NHC forecast advisory 28, which was issued 2012/08/28/0300 UTC

An example output product is shown in Fig. 5, in which the maximum significant wave heights for NHC forecast advisory 28 are visualized within a Google Earth application. This advisory was issued at 2012/08/28/0300 UTC, or less than 24 h before the storm's initial landfall within the Mississippi River Delta. Isaac was centered within the Gulf at (87.0°W, 27.1°N), or due south from Pensacola and due west from Cape Coral, Florida. As the storm moved along its predicted track toward Louisiana, the SWAN-computed waves had significant heights with maxima of 9–10 m along the storm track and 5 m or larger throughout a large section of the northern Gulf. These forecast products were shared in real time with the NWS Forecast Offices in Tallahassee and Miami.

The variability in official forecast tracks and storm intensities caused significant differences in the ADCIRC-computed surge across the forecasts, including NHC forecast advisories 20, 24, and 28 in the days leading to landfall. Advisory 20 (Fig. 6a) was the last forecast with a predicted landfall location in the Florida panhandle. Had the storm followed this track, Louisiana would have been located on its weaker western side, and thus the predicted surge levels were minimal along its coastline. By Advisory 24 (Fig. 6b), the predicted landfall location had moved to Louisiana. However, at this time, the storm had just moved into the Gulf, and it was expected to intensify more than actually happened. For this reason, the predicted surge levels were larger than 4 m in Lake Borgne east of New Orleans.

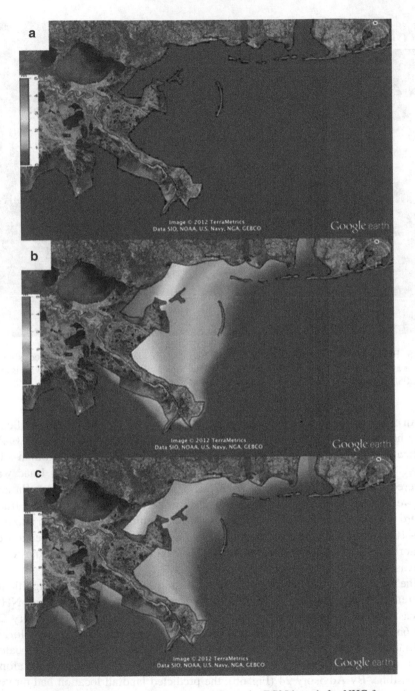

Fig. 6 Maximum ADCIRC-computed water levels on the EC95d mesh for NHC forecast advisories (**a**) 20, issued 2012/08/26/0300 UTC and with a predicted landfall in the Florida panhandle; (**b**) 24, issued 2012/08/27/0300 UTC and with a predicted landfall near Grand Isle; and (**c**) 28, issued 2012/08/28/0300 UTC, or less than 24 h before the initial landfall. The contour range is the same in all three panels and has a maximum of 5 m

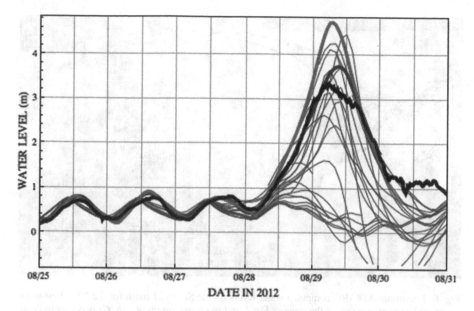

Fig. 7 Hydrographs at NOAA station 8761305 near Shell Beach, Louisiana. Water levels are shown in different colors for the measurements (*black*), EC95d forecasts (*gray*), EC95d hindcast (*red*) and SL16v31 hindcast (*blue*). ADCIRC-computed water levels have been adjusted to the MLLW datum used by the NOAA measurements. Colors refer to figure as printed online

When advisory 28 (Fig. 6c) was issued the next day, the predicted storm intensities lessened, and the maximum surge was predicted to reach a maximum of about 3 m in Lake Borgne.

This forecast guidance was reasonable, especially considering the relatively modest resources used to compute it. Figure 7 shows the time series of measured and computed water levels at the NOAA station 8761305 near Shell Beach, Louisiana. This station is located on the southern shoreline of Lake Borgne just east of metropolitan New Orleans, and it experienced the large surges that threatened the protection system surrounding the city and its neighboring communities. The peak surge was measured at 3.4 m at this location. It should be noted that the NOAA station near Shell Beach lies approximately 22 km from the western side of Lake Borgne, which experienced the highest surge levels during Isaac. The predicted surges from the EC95d forecasts show a range of peak values, from about 1.5 m to about 4.4 m, reflecting the uncertainty in the landfall location and intensity as the storm evolved.

Thus, the SWAN+ADCIRC wave and circulation models were employed within the ASGS to provide forecast guidance. The predicted maxima for significant wave heights and water levels were visualized with FigureGen and shared as KMZ files for use within Google Earth, allowing the user to control the zoom level, overlay other forecast products, add user-defined layers, etc. However, this guidance was limited to the open water, since the coarsely resolved EC95d mesh does not represent the coastal floodplains of southeastern Louisiana.

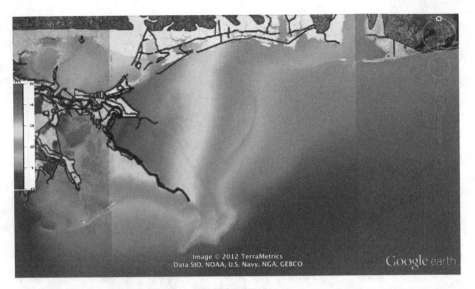

Fig. 8 Maximum ADCIRC-computed water levels on the SL16v31 mesh for the NHC best-track analysis. The contour range is the same as Fig. 6 and has a maximum of 5 m. *Gray colors* indicate regions that were not wetted within the ADCIRC simulation. Colors refer to figure as printed online

4.2.2 Hindcast on the SL16v31 Mesh

After the event, the storm parameters within the NHC best-track analysis were used to create an asymmetric wind field using the same methodology as that used during the forecasts [20, 27], and then hindcasts were performed on the EC95d and SL16v31 meshes. The EC95d hindcast produced too much surge on the Louisiana–Mississippi continental shelf and within the marshes to the east of New Orleans. At the Shell Beach station (red line in Fig. 7), the peak surge is about 4.75 m, and thus larger than the peak surges from the measurements or any of the forecasts.

Much of the error can be attributed to the coarse resolution of the EC95d mesh. The computed surge is appropriately smaller in the SL16v31 hindcast, including by as much as 1 m smaller at the Shell Beach station (blue line in Fig. 7). The increased resolution in the SL16v31 mesh allows for a better representation of the continental shelf, the wave-breaking zones, the coastal floodplains, and the natural and manmade channels that convey surge into the inland lakes and estuaries. This behavior is evident in Fig. 8, which shows the maximum ADCIRC-computed water levels for the SL16v31 hindcast. Surge is pushed into the Biloxi and Caernarvon marshes and against the earthen levees near Braithwaite. Compared to the results from EC95d, which limits the surge to the coastline, these high-resolution hindcast results are a better representation of the flooding caused by Isaac.

These Isaac hindcasts can be further improved through the refinement of wind fields with assimilated measurements of atmospheric pressures and wind speeds. Detailed validation studies are forthcoming, and they will examine the forecast performance on the high-resolution meshes utilized by the LSU/UNC team during the event, as well as the differences in SWAN+ADCIRC responses when forced with a variety of forecast winds.

5 Conclusions

Hurricane forecasts require hydrodynamic models that represent the generation of waves and storm surge and their propagation toward the coastline. These models should be employed on computational meshes with appropriate resolution to represent details of the complex coastal environment, including the coastal floodplains and fine-scale channels. During the 2012 hurricane season, the tightly coupled SWAN+ADCIRC models were employed within the ASGS to provide forecast guidance on several meshes for delivery to several stakeholders. These forecasts must be visualized efficiently and shared in formats that geo-reference the model results.

The FigureGen postprocessing tool was developed for visualization of simulation files from SWAN+ADCIRC. FigureGen visualizes input data such as bathymetry and vertex-based attributes such as Manning's n values and directional wind reduction factors, as well as computed output data such as significant wave heights and water levels. Scalar and vector data can be plotted with filled and/or linear contours, and vector data can also be overlaid on a regular grid. In a parallel computing environment, FigureGen uses MPI to divide the work, and the accelerated workflow is limited only by the number of computational cores. Illustrations can be geo-referenced for use in other geospatial software.

FigureGen was incorporated within the ASGS, and it was used to visualize forecast guidance during Hurricane Isaac (2012). Uncertainties in the track and intensity caused significant differences in the forecasts as the storm approached the northern Gulf coastline. On a coarsely resolved EC95d mesh, the computed peak surges ranged from 1.5 m to 4.4 m at a NOAA station east of New Orleans that had a measured peak value of about 3.4 m. However, when the storm was hindcasted using a high-resolution SL16v31 mesh and winds generated from the NHC best-track analysis, the computed peak surge was within 0.5 m of the field measurements coinciding with the actual landfall of Isaac. Future work will explore the ASGS forecast performance on higher-resolution meshes.

The forecast guidance was visualized in KMZ format for use within Google Earth, allowing the user to control the zoom level, overlay additional forecast products, and add user-defined layers. As Isaac crossed the northwestern Caribbean and Gulf of Mexico, the results of the ASGS forecasts were relayed as visualization products (KMZ and PNG files) to the NWS Southern Region Headquarters and NWS forecast offices in Miami and Tallahassee. In the Texas SOC, the ASGS forecast products were used to brief the Governor's Emergency Management Council, and descriptions of the results appeared in situation reports published by the Texas Division of Emergency Management. The rapid delivery of updated information about potential storm impacts following the release of each NHC forecast advisory offered an excellent new source of accurate guidance to weather forecasters and emergency managers in the path of Isaac. Future work will tighten the coupling between FigureGen and SWAN+ADCIRC. Model results will be visualized as they become available, and thus the illustrations will be ready for the user when the simulation finishes.

Acknowledgements This work was supported by awards from the National Science Foundation (DMS-0915223); the SSPEED Center at Rice University (http://sspeed.rice.edu/); the Gulf of Mexico Research Initiative (http://gulfresearchinitiative.org/); and the Coastal Hazards Center of Excellence, a U.S. Department of Homeland Security Science and Technology Center of Excellence (2008-ST-061-ND 0001). Computational resources were provided by the Texas Advanced Computing Center (http://www.tacc.utexas.edu/) and the Extreme Science and Engineering Discovery Environment (under award number TG-080016N). Some images were overlaid on the ocean basemap that was designed and developed by Esri [11].

References

1. Arcement, G.J., and Schneider, V.R. Guide for selecting Manning's roughness coefficients for natural channels and flood plains. U.S. Geological Survey Water Supply Paper 2339, U.S. Geological Survey, Denver, CO, 38pp. (1989).
2. Barnes, H.H. Roughness characteristics of natural channels. U.S. Geological Survey Water Supply Paper 1849, U.S. Geological Survey, Washington, DC, 213pp. (1967).
3. Blanton, B.O., McGee, J., Fleming, J.G., Kaiser, C., Kaiser, H., Lander, H., Luettich Jr., R.A., Dresback, K.M., and Kolar, R.L. Urgent computing of storm surge for North Carolina's coast. Proceedings of the International Conference on Computational Science, ICCS 2012, Procedia Computer Science, 9, 1677–1686 (2012).
4. Booij, N., and Holthuijsen, L.H. Propagation of ocean waves in discrete spectral wave models. Journal of Computational Physics, 68, 307–326 (1987).
5. Booij, N., Ris, R.C., and Holthuijsen, L.H. A third-generation wave model for coastal regions, Part I, Model description and validation. Journal of Geophysical Research, 104, 7649–7666 (1999).
6. Brown, D., and Brennan, M. (2012). Hurricane Isaac Tropical Cyclone Position Estimate. National Hurricane Center, August 28, http://www.nhc.noaa.gov/archive/2012/al09/al092012.posest.08282356.shtml.
7. Chow, V.T. Open-Channel Hydraulics. McGraw-Hill Book Company, 680pp. (1959).
8. Bunya, S., Dietrich, J.C.,Westerink, J.J., Ebersole, B.A., Smith, J.M., Atkinson, J.H., Jensen, R.E., Resio, D.T., Luettich Jr., R.A., Dawson, C.N., Cardone, V.J., Cox, A.T., Powell, M.D., Westerink, H.J., and Roberts, H.J. A High Resolution Coupled Riverine Flow, Tide, Wind, WindWave and Storm Surge Model for Southern Louisiana and Mississippi: Part I: Model Development and Validation. Monthly Weather Review, 138(2), 345–377 (2010).
9. Dawson, C.N., Westerink, J.J., Feyen, J.C., and Pothina, D. Continuous, Discontinuous and Coupled Discontinuous–Continuous Galerkin Finite Element Methods for the Shallow Water Equations. International Journal for Numerical Methods in Fluids, 52, 63–88 (2006).
10. Dawson, C.N., Kubatko, E.J., Westerink, J.J., Trahan, C.J., Mirabito, C., Michoski, C., and Panda, N. Discontinuous Galerkin Methods for Modeling Hurricane Storm Surge. Advances in Water Resources, DOI 10.1016/j.advwatres.2010.11.004, 34, 1165–1176 (2011).
11. DeMeritt, M. A Foundation for Ocean GIS. ArcUser, Fall 2011.
12. Dietrich, J.C., Kolar, R.L., Luettich Jr, R.A. Assessment of ADCIRCs Wetting and Drying Algorithm. Proceedings of Computational Methods in Water Resources, C.T. Miller, M.W. Farthing, W.G. Gray, and G.F. Pinder, eds., 2, 1767—1778 (2004).

13. Dietrich, J.C., Bunya, S., Westerink, J.J., Ebersole, B.A., Smith, J.M., Atkinson, J.H., Jensen, R.E., Resio, D.T., Luettich Jr., R.A., Dawson, C.N., Cardone, V.J., Cox, A.T., Powell, M.D.,Westerink, H.J., and Roberts, H.J. A High Resolution Coupled Riverine Flow, Tide, Wind, Wind Wave and Storm Surge Model for Southern Louisiana and Mississippi: Part II: Synoptic Description and Analyses of Hurricanes Katrina and Rita. Monthly Weather Review, 138, 378–404 (2010).
14. Dietrich, J.C., Zijlema, M., Westerink, J.J., Holthuijsen, L.H., Dawson, C.N., Luettich Jr, R.A., Jensen, R.E., Smith, J.M., Stelling, G.S., Stone, G.W. Modeling Hurricane Waves and Storm Surge using Integrally-Coupled, Scalable Computations. Coastal Engineering, 58, 45–65, DOI:10.1016/j.coastaleng.2010.08.001 (2011a).
15. Dietrich, J.C., Westerink, J.J., Kennedy, A.B., Smith, J.M., Jensen, R.E., Zijlema, M., Holthuijsen, L.H., Dawson, C.N., Luettich Jr., R.A., Powell, M.D., Cardone, V.J., Cox, A.T., Stone, G.W., Pourtaheri, H., Hope, M.E., Tanaka, S., Westerink, L.G., Westerink, H.J., and Cobell, Z. Hurricane Gustav (2008) Waves and Storm Surge: Hindcast, Validation and Synoptic Analysis in Southern Louisiana. Monthly Weather Review, 139(8), 2488–2522 (2011b).
16. Dietrich, J.C., Trahan, C.J., Howard, M.T., Fleming, J.G., Weaver, R.J., Tanaka, S., Yu, L., Luettich Jr, R.A., Dawson, C.N., Westerink, J.J., Wells, G., Lu, A., Vega, K., Kubach, A., Dresback, K.M., Kolar, R.L., Kaiser, C., Twilley, R.R. (2012). Surface Trajectories of Oil Transport along the Northern Coastline of the Gulf of Mexico. Continental Shelf Research, 41(1), 17–47, DOI:10.1016/j.csr.2012.03.015 (2012a).
17. Dietrich, J.C., Tanaka, S., Westerink, J.J., Dawson, C.N., Luettich Jr, R.A., Zijlema, M., Holthuijsen, L.H., Smith, J.M., Westerink, L.G., Westerink, H.J. Performance of the Unstructured-Mesh, SWAN+ADCIRC Model in Computing Hurricane Waves and Surge. Journal of Scientific Computing, 52(2), 468–497, DOI:10.1007/s10915-011-9555-6 (2012b).
18. Fleming, J.G., Fulcher, C., Luettich Jr., R.A., Estrade, B., Allen, G., and Winer, H. A Real Time Storm Surge Forecasting System using ADCIRC. Proceedings of Estuarine and Coastal Modeling X, Spaulding, M.L. (ed.), ASCE, 373–392 (2008).
19. Garratt, J.R. Review of drag coefficients over oceans and continents. Monthly Weather Review, 105, 915–929 (1977).
20. Holland, G.J. An analytical model of the wind and pressure proles in hurricanes. Monthly Weather Review, 108, 1212–1218 (1980).
21. Karypis, G., and Kumar, V. A fast and high quality multilevel scheme for partitioning irregular graphs. SIAM Journal of Scientific Computing, 20(1), 359–392 (1999).
22. Kennedy, A.B., Gravois, U., Zachry, B.C., Westerink, J.J., Hope, M.E., Dietrich, J.C., Powell, M.D., Cox, A.T., Luettich Jr., R.A., and Dean, R.G.. Origin of the Hurricane Ike Forerunner Surge. Geophysical Research Letters, 38, L08608, DOI 10.1029/2011GL047090 (2011).
23. Kolar, R.L., Westerink, J.J., Cantekin, M.E., and Blain, C.A. Aspects of nonlinear simulations using shallow water models based on the wave continuity equations. Computers and Fluids, 23(3), 1–24 (1994).
24. Le Provost, C., Lyard, F., Molines, J., Genco, M., and Rabilloud, F. A hydrodynamic ocean tide model improved by assimilating a satellite altimeter-derived data set. Journal of Geophysical Research, 103, 5513–5529, (1998).
25. Luettich Jr., R.A., and Westerink, J.J. Formulation and Numerical Implementation of the 2D/3D ADCIRC Finite Element Model Version 44.XX, http://adcirc.org/adcirc_theory_2004_12_08.pdf (2004).
26. Martyr, R.C., Dietrich, J.C., Westerink, J.J., Kerr, P.C., Dawson, C.N., Smith, J.M., Pourtaheri, H., Powell, N., van Ledden, M., Tanaka, S., Roberts, H.J., Westerink, L.G., and Westerink, H.J. Simulating Hurricane Storm Surge in the Lower Mississippi River under Varying Flow Conditions. Journal of Hydraulic Engineering, 139(5), 492–501, DOI: 10.1061/(ASCE)HY.1943-7900.0000699 (2013).
27. Mattocks, C., and Forbes, C. A real-time, event-triggered storm surge forecasting system for the state of North Carolina. Ocean Modelling, 25, 95–119 (2008).
28. Mukai, A., Westerink, J.J., Luettich Jr., R.A., and Mark, D. Eastcoast 2001: A tidal constituent database for the Western North Atlantic, Gulf of Mexico and Caribbean Sea. Technical Report ERDC/CHL TR-02-24, U.S. Army Corps of Engineers, 201pp., (2002).

29. Pasch, R., and Roberts, D. (2012). Hurricane Isaac Tropical Cyclone Position Estimate. National Hurricane Center, August 28, http://www.nhc.noaa.gov/archive/2012/al09/al092012. posest.08290400.shtml.
30. Pasch, R., and Roberts, J. (2012). Hurricane Isaac Tropical Cyclone Position Estimate. National Hurricane Center, August 29, http://www.nhc.noaa.gov/archive/2012/al09/al092012. posest.08290758.shtml.
31. Powell, M.D., Vickery, P.J., and Reinhold, T.A. Reduced drag coefficient for high wind speeds in tropical cyclones. Nature, 422, March 20, 279–283 (2003).
32. Powell, M.D. Drag Coefficient Distribution and Wind Speed Dependence in Tropical Cyclones. Final Report to the National Oceanic and Atmospheric Administration (NOAA) Joint Hurricane Testbed (JHT) Program (2006).
33. Seahorse Coastal Consulting (2012a). The ASGS Developer's Guide 2011. Available online at: http://www.seahorsecoastal.com/ASGSDevGuide2011.pdf.
34. Seahorse Coastal Consulting (2012b). The ASGS Operator's Guide 2011. Available online at: http://www.seahorsecoastal.com/ASGSOperatorsGuide2011.pdf.
35. Stelling, G.S., and Leendertse, J.J. Approximation of convective processes by cyclic AOI methods. Proceedings of the 2nd international conference on estuarine and coastal modeling, ASCE Tampa, Florida, 771–782 (1992).
36. Tanaka, S., Bunya, S., Westerink, J.J., Dawson, C.N., and Luettich Jr., R.A. Scalability of an Unstructured Grid Continuous Galerkin Based Hurricane Storm Surge Model. Journal of Scientific Computing, 46, 329–358 (2011).
37. Wessel, P., and Smith, W.H.F. Free software helps map and display data, EOS Trans. AGU, 72, 441 (1991).
38. Westerink, J. J., Luettich Jr., R.A., and Scheffner, N.W. ADCIRC: An advanced three-dimensional circulation model for shelves, coasts, and estuaries, Report 3: Development of a tidal constituent database for the western North Atlantic and Gulf of Mexico. Technical Report DRP 92-6, U.S. Army Engineer Research and Development Center, Coastal and Hydraulics Laboratory, Vicksburg, MS (1993).
39. Westerink, J.J., Luettich Jr., R.A., and Muccino, J.C. Modeling tides in the western North Atlantic using unstructured graded grids. Tellus 46A, 178–199 (1994).
40. Westerink, J.J., Luettich Jr., R.A., Feyen, J.C., Atkinson, J.H., Dawson, C.N., Roberts, H.J., Powell, M.D., Dunion, J.P., Kubatko, E.J., Pourtaheri, H. A Basin to Channel Scale Unstructured Grid Hurricane Storm Surge Model Applied to Southern Louisiana. Monthly Weather Review, 136, 3, 833–864 (2008).
41. Zijlema, M. Computation of wind-wave spectra in coastal waters with SWAN on unstructured grids. Coastal Engineering, 57, 267–277 (2010).

Methane in Subsurface: Mathematical Modeling and Computational Challenges

Malgorzata Peszynska

Abstract We discuss mathematical and computational models of two applications important for global climate and energy recovery involving the evolution of methane gas in the subsurface. In particular, we develop advanced models of adsorption occurring in coalbed methane recovery processes, and discuss the underlying conservation laws with nonstandard terms. Next we describe the phase transitions relevant for modeling methane hydrates in subsea sediments, where the major challenge comes from implementation of solubility constraints. For both applications we formulate the discretization schemes and outline the main challenges in convergence analysis and solver techniques. We also motivate the need for continuum and discrete models at porescale.

Keywords Multiphase and multicomponent flow and transport • Porous media • Adsorption models • Phase transitions • Finite differences • Methane hydrates • Semismooth functions • Semismooth Newton methods • Hysteresis • Coalbed methane • Mean-field equilibrium models

MSC Classification (2010): 76S05, 65M06, 65M22, 74N30, 76V05, 80A30, 65M75

1 Introduction

Methane is both a greenhouse gas and an energy resource. In this paper we discuss the challenges in computational modeling of methane in two applications important

M. Peszynska (✉)
Department of Mathematics, Oregon State University, Corvallis, OR 97331, USA
e-mail: mpesz@math.oregonstate.edu

C. Dawson and M. Gerritsen (eds.), *Computational Challenges in the Geosciences*, The IMA Volumes in Mathematics and its Applications 156, DOI 10.1007/978-1-4614-7434-0_4, © Springer Science+Business Media New York 2013

for global climate and energy studies, namely enhanced coalbed methane (ECBM) recovery, and modeling methane hydrate evolution in subsea sediments (MH).

Coalbed methane is a form of natural gas extracted from coal beds. In recent decades it has become an important source of energy in the United States and other countries, and coal and methane are important energy resources exported from the United States. In the ECBM technology, carbon dioxide and/or nitrogen or other gases are injected into unmineable coal seams to promote displacement and extraction of methane. Recent pilot projects in various countries evaluated ECBM as a potential carbon sequestration technology [46,49,51,82,135,140,148]. The technology appears promising but is associated with various uncertainties and hazards, not the least of which include incomplete understanding of the underlying processes and difficulties with carrying out experiments.

Methane hydrates (MH), an icelike substance containing methane molecules trapped in a lattice of water molecules, are present in large amounts along continental slopes and in permafrost regions and therefore, are a possible source of energy [3, 47, 86, 129], and at the same time a potential environmental hazard [29, 33, 38, 57, 127]. There are recent initiatives by the Department of Energy's (DOE) National Energy Technology Laboratory (NETL), in collaboration with the U.S. Geological Survey (USGS), and an industry consortium led by Chevron, in gas hydrate drilling, research expeditions [6], and observatories [5, 7] that help to evaluate methane hydrate as an energy resource. Although the existence of gas hydrates in nature has been known for many decades, our understanding of their potential impact on slope stability, the biosphere, carbon cycling, and climate change is still evolving.

In this paper we give an overview of the many open mathematical and computational questions that arise for these applications. They are relatively unknown to the mathematical and numerical communities. We first consider their traditional continuum models, which are systems of partial differential equations (PDEs) at mesoscale, i.e., lab or field scale. See Sect. 2 for notation, and Sects. 3 and 4 for ECBM and MH models, respectively. Since these PDEs are nonlinear and coupled, no general theory of well-posedness or of convergence of approximation schemes is available. However, some analyses can be pursued for submodels of ECBM and MH, and their structure calls for a family of related mathematical and computational techniques and solvers. For the latter, some techniques originally developed for optimization are emerging as effective methods for solving nonlinear problems with inequality constraints and piecewise smooth nonlinearities.

The accuracy of computational models depends on the ability of the physical models themselves to describe all relevant phenomena and on the precision of their data. One common theme for ECBM and MH models is that they describe metastable phenomena such as adsorption and phase transitions whose physics may be either poorly understood or difficult to capture at mesoscale. In addition, these evolution phenomena affect the porescale, which in turn changes the flow and transport characteristics. In Sect. 5 we propose some models at porescale that can help to overcome the limitations of continuum modeling and provide a lookup library for the missing experimental data. In the future, these models can be combined with the continuum models in static or dynamic hybrid schemes.

2 Processes and Continuum Models

Models of methane evolution must account for multiple phases and multiple components evolving in time in a porous reservoir $\Omega \subset \mathbb{R}^d$, $1 \leq d \leq 3$, under the Earth's surface at depth $D(\mathbf{x})$, $\mathbf{x} \in \Omega$, and porosity ϕ and permeability K. Recall that ϕ is a positive scalar, and K is a uniformly positive definite tensor. Also, let P, T denote pressure and temperature. Multiphase multicomponent flow and transport models are generally coupled systems of nonlinear PDEs with additional volume, capillary, and thermodynamic constraints; see standard developments in [28,45,67].

The mass conservation equations written for each component C read as follows:

$$\overset{storage}{\overbrace{\mathbf{Sto}_C}} + \overset{advection}{\overbrace{\mathbf{Adv}_C}} + \overset{diffusion}{\overbrace{\mathbf{Diff}_C}} = \overset{source}{\overbrace{q_C}}, \text{ in } \Omega, \tag{2.1}$$

in which the individual terms depend on the component C under study, on the process, and on its scale. Alternatively, one can write such equations for each component C and each phase p, and account for the transfer of mass of a component between phases. Data for these dynamics are hard to obtain experimentally; however, they can be provided by porescale studies; see Sect. 5.

The definition of $\mathbf{Sto}_C$ is crucial for each application, since it specifies how a component moves out of one form or phase and whether the phase transitions or adsorption takes place in equilibrium or via a kinetic model. The $\mathbf{Sto}_C$ terms include the time rate of change of all forms of the component C dissolved in various flowing liquid or gas phases, present in a stationary phase such as adsorbed on a surface of porous skeleton or part of a gas-hydrate crystal. In other words,

$$\mathbf{Sto}_C := \frac{\partial}{\partial t}(\phi N_C) := \frac{\partial}{\partial t}(\phi \sum_p \rho_p S_p \chi_{pC}).$$

Here we used the total mass concentration of a component N_C, with S_p denoting the saturations/volume fractions, and ρ_p the density of phase p, respectively. We have $S_p \geq 0$ and $\sum_p S_p = 1$, and densities are given from an appropriate equation of state. Also, we used mass fractions of a component C in phase p, denoted by χ_{pC}. We have $\chi_{pC} \geq 0$, and for each phase, $\sum_C \chi_{pC} = 1$.

Next we discuss the terms $\mathbf{Adv}$ and $\mathbf{Diff}$. Their definitions depend on the application and scale. Typically, the diffusion/dispersion terms $\mathbf{Diff}$ are the divergence of diffusive fluxes formulated using Fick's law

$$\mathbf{Diff}_C := \nabla \cdot \left(\phi \sum_p \rho_p S_p D_{pC} \nabla \chi_{pC} \right), \tag{2.2}$$

and $\mathbf{Adv}$ includes divergence of mass fluxes

$$U_p = -K \frac{k_{rp}}{\mu_p}(\nabla P_p - \rho_p G \nabla D(\mathbf{x})), \quad p = l, g, \tag{2.3}$$

$$\mathbf{Adv}_C := \nabla \cdot \sum_p \chi_{pC} \rho_p U_p, \tag{2.4}$$

with the velocities given via the multiphase extension of Darcy's law, where k_{rp}, μ_p denote relative permeability and viscosity of phase p, and G the gravitational constant. In addition, the phase pressures of mobile phases P_l, P_g are coupled via the capillary pressure relationship $P_g - P_l = P_c(S_l)$ given by Brooks–Corey relationships or van Genuchten correlations [28, 56, 67].

The model (2.1) is very general; difficulties in its analysis and approximation arise from the presence of multiple components as well as from the various nonlinear couplings. Additionally, the data for some of these couplings are difficult to obtain; see Sect. 5. Special modeling constructs can be used to decrease the computational complexity associated, e.g., with multiple scales, nonequilibrium conditions, and metastability. On the other hand, these require additional care in numerical approximation.

We illustrate some of the relevant issues in a discussion of scalar (one-component) equations in $d = 1$ to follow. Throughout, we rescale the variables and eliminate constants, for example ϕ, to emphasize the qualitative rather than quantitative structure of the models.

First we consider similarities and differences in the generic scalar PDEs for ECBM and MH. In Sects. 3 and 4 we elaborate on their details.

In the single-phase rescaled scalar version of (2.1) for ECBM we have

$$\frac{\partial}{\partial t}(\chi + \Upsilon) + \mathbf{Adv} + \mathbf{Diff} = 0, \tag{2.5}$$

where χ denotes the concentration or mass fraction of the mobile component, and Υ is that of the adsorbed amount. The connection between χ, Υ and/or their rates has to be provided via an equilibrium, kinetic, or hysteretic relationship

$$(\chi, \Upsilon) \in F. \tag{2.6}$$

Here F is a single- or multivalued stationary or evolution relationship related to the graph of an adsorption/desorption isotherm.

For MH, the evolution of the methane component is governed, in a simplified rescaled version, by

$$\frac{\partial}{\partial t}((1 - \Upsilon)\chi + R\Upsilon) + \mathbf{Adv} + \mathbf{Diff} = 0. \tag{2.7}$$

Here (χ, Υ) are methane solubility and hydrate or gas saturation, respectively, and R is a constant. The variables (χ, Υ) are bound together by the volume and solubility constraints, which can be written as (2.6).

The discretization of the models (2.5), (2.7) depends on the definition of **Adv, Diff** and their relative significance. For example, for dominant diffusion, finite elements in space and implicit backward Euler schemes in time can be used. For dominant advection, conservative consistent finite difference schemes are needed, and the advection terms are typically handled explicitly in time. When both diffusion and advection are present we can use an operator splitting procedure.

For a simple F in (2.6), fairly well known results on the discretization and analyses are available. For complicated F, the handling of (2.6), especially for multivalued relations and analysis, is a challenge. In Sects. 3 and 4 we provide details relevant to ECBM and MH; we focus on the representation and approximation of F and discuss the associated stability and solver issues. In Sect. 5 we propose the use of hybrid computational models that can be used to obtain data for F and other essential model ingredients.

Finally, the initial and boundary conditions need to be specified to close any system of PDEs and to discuss its well-posedness. However, analysis of either ECBM or MH has not been attempted, and coupled nonlinear systems of partial differential equations of mixed, variable, and degenerate type do not have a general well-posedness theory except for special submodels [10, 11, 25, 98, 125]. The lack of a general theory makes the numerical approximation delicate; it is necessary to thoroughly understand each subproblem as well as the challenges of the coupled system, to the extent possible, before a particular numerical discretization method is selected and its results used. Various simulators have been successfully implemented for typical multiphase multicomponent models, and these are based on carefully selected spatial discretization schemes and nonlinear solvers [2,4,45,70,77,97,104, 110, 138]. In the exposition below we discuss avenues for possible analyses and algorithmic extensions of some of these schemes.

3   Transport with Adsorption in ECBM

Coal beds have the form of coal *seams* surrounded by sandstone, gravel, mudstone, or shale. The coal seams have a multiscale structure of microporous coal matrix interspersed with *cleats*, i.e., fractures or macropores.

The majority of transport occurs in the cleats accompanying the flow of gas and possibly of water, while the majority of storage occurs in the matrix where gases undergo diffusion and adsorption, close to supercritical conditions. For ECBM, the components are $C = M, D, N, (W)$ methane, carbon dioxide, nitrogen, and (for wet gas models) water. The phases in which these components can remain are $p = g, a, l$: gas, adsorbed gas, and liquid phase (for wet gas).

In a simple dry-gas model with $S_g = 1$ for ECBM with $C = M, D, N$, with $p = g, a$ we have

$$\chi_{gM} + \chi_{gD} + \chi_{gN} = 1, \tag{3.1}$$

$$\chi_{aM} + \chi_{aD} + \chi_{aN} = 1, \tag{3.2}$$

so that equations (2.1) are written for the evolution of N_M, N_D, N_N.

When gases such as carbon dioxide are injected into coal seams, they make their way through the cleats into the micropore structure of the matrix. Here they preferentially adsorb, displacing methane from adsorption sites; subsequently, this methane is transported through cleats and is available for extraction. The predominant transport mechanism in cleats (macropores) is that of advection, while the transport into and out of the coal matrix occurs through diffusion into mesopores and micropores, where the gases undergo adsorption and desorption at the surface of the grains.

Adsorption is a surface phenomenon in which particles, molecules, or atoms of adsorbate attach to the surface of adsorbent. It has numerous technical applications ranging from water purification, chromatography, to drug delivery, and is fundamental in ECBM. There is current research on the nature and occurrence of adsorption–desorption hysteresis [27, 112, 122] and competitive, or preferential, adsorption [21, 23, 24, 55, 62, 78, 123]. Functional relationships that fit some of the experimental data have been proposed in [20, 26, 27, 31, 81, 91, 126], but comprehensive models are lacking.

3.1 Adsorption Models

Consider now a relationship between $\chi := \chi_{gM}$ and $\Upsilon := \chi_{aM}$ that needs to be provided to complement (3.1), (3.2) for the methane component $C = M$. (Similar relationships are needed for $C = D, N$). Basic experimental models are usually given as equilibrium isotherms [1, 39] such as

$$\Upsilon = f(\chi), \tag{3.3}$$

where f is a known smooth monotonically increasing function that describes the surface coverage Υ of adsorbent depending on the gas or vapor pressure χ of adsorbate, at constant temperature, in equilibrium. For example, the well-known Langmuir type-I isotherm [1] has the form

$$\Upsilon = f(\chi) = \Upsilon^{\max} \frac{b\chi}{1 + b\chi}, \tag{3.4}$$

where $b, \Upsilon^{\max}$ are constants. It is derived from the monolayer assumption and equality of adsorption and desorption rates. This isotherm applies well in a variety of microporous, mesoporous, and macroporous media in sub-, near-, and supercritical conditions [39, 62]. Note that $\Upsilon = f(\chi)$ is smooth, concave, increasing, and Lipschitz.

An alternative to the equilibrium model (3.3) is the kinetic model

$$\frac{d\Upsilon}{dt} = r(f(\chi) - \Upsilon), \tag{3.5}$$

in which Υ is an exponential follower (with rate r) of the equilibrium model $f(\chi)$. See [14–16, 35, 36], where numerical schemes for (2.5) and (3.3) or (3.5) were proposed and analyzed. Difficulties arise if the isotherm is not Lipschitz in (3.3)–(3.5), but this case is not relevant for ECBM. See also [133] for a discussion of stability of the Godunov (upwind) method for the equilibrium and kinetic models, as well as their relationship in $d = 1$.

We discuss now the model for **Diff** ~ 0, **Adv** $= \nabla \cdot \chi$. Then (2.5) is of hyperbolic type and can be rewritten as a scalar conservation law via a change of variable $w = \chi + \Upsilon$,

$$w_t + \nabla \cdot g(w) = 0, \tag{3.6}$$

with an increasing and convex flux function $g \sim (I + f)^{-1}$. It is well known [71] that its solutions develop singularities in finite time from smooth initial data. An appropriate numerical method needs to be conservative and stable. For example, this is satisfied by the Godunov method, written here in primary variables for (2.5),

$$\chi_j^n + \Upsilon_j^n + \lambda (\chi_j^{n-1} - \chi_{j-1}^{n-1}) = \chi_j^{n-1} + \Upsilon_j^{n-1}, \tag{3.7}$$

where $\lambda = k/h$, k is the time step, and h is the spatial discretization parameter. The discretization is defined at the discrete spatial and temporal points $x_j = jh, t_n = nk$. The scheme is stable, provided $\lambda \le 1 + f'(\chi_j^n)$, which is easy to satisfy if, e.g., $\lambda \le 1$.

Strictly speaking, a Godunov scheme for (3.6) is entirely explicit and does not require a solution of any global linear or nonlinear system. However, the scheme in (3.7) is not, since it requires solving

$$\chi_j^n + \Upsilon_j^n = \chi_j^n + f(\chi_j^n) = A_j, \tag{3.8}$$

where $A_j := -\lambda (\chi_j^{n-1} - \chi_{j-1}^{n-1}) + \chi_j^{n-1} + \Upsilon_j^{n-1}$ is known from the previous time step, and where we substituted (3.3) into (2.5).

Solving (3.8) has to be done only locally at every grid point. In fact, for some isotherms, e.g., Langmuir, the algebraic form of (3.8) is very simple, and (3.8) can be solved explicitly for χ_j^n. In other cases, a simple local Newton iteration suffices, since $I + f$ is a smooth bijective function. To avoid unphysical negative iterates for χ_j^n, it is sufficient to use an initial guess determined from the zero of the linear model of f at 0, i.e., of $f(\chi) \approx \Upsilon^{max} b\chi$.

Now consider the nonequilibrium case, i.e., the case in which (2.5) is complemented by (3.5). Formally, we have to solve now, in addition to (3.7), a coupled ODE defined at every grid point

$$\Upsilon_j^n = \Upsilon_j^{n-1} + kr \left(f(\chi_j^{\tilde{n}}) - \Upsilon_j^{\tilde{n}} \right), \tag{3.9}$$

where $\bar{n}, \tilde{n}$ denote either n or $n-1$. For small r and moderate f' one can find a sufficiently small k to ensure the conditional stability of the explicit solution. For large r the system is stiff and therefore calls for implicit treatment with $\bar{n} = \tilde{n} = n$, so that

$$\chi_j^n + \Upsilon_j^n = A_j, \tag{3.10}$$

$$-krf(\eta_j^n) + (1+kr)\Upsilon_j^n = B_j = \Upsilon_j^{n-1}, \tag{3.11}$$

where A_j, B_j are known from the previous time step. Its solvability for a monotone f is analyzed similarly to (3.8).

In what follows we first discuss the extensions to the basic scalar models defined above, and next we discuss the multicomponent case.

3.2 Diffusion into Micropores and Transport with Memory Terms

In ECBM, the predominantly advective transport in cleats is accompanied by a range of diffusive phenomena in which the molecules of methane and carbon dioxide migrate and get adsorbed to the meso- and micropores of the coal. This is known to affect the molecular structure of the coal matrix and leads to experimentally observed phenomena such as coal matrix swelling, which has a distinct kinetic character [24, 78, 90, 123, 137, 150].

Micropore and mesopore diffusion has been included in the classical bidisperse model [65, 114] extended to include adsorption in micropores in [26] and to realistic gas transport models [124], with ECBM-related experimental work on the rates of kinetics in [20]. See also [26, 65, 114, 123].

These models include in $\mathbf{Sto}_C$ the quantity Υ living in micropores, which is governed by its own diffusion equation at a lower scale. Thus, Υ is related to χ via a convolution

$$\frac{\partial \Upsilon}{\partial t} := \frac{\partial \chi}{\partial t} * \beta := \int_0^t \frac{\partial \chi}{\partial t}(t-s)\beta(s)\mathrm{d}s. \tag{3.12}$$

Here $\beta \in C^1(0,T)$ is a weakly singular or bounded monotone decreasing kernel, i.e., it is locally integrable, with $\beta' \le 0$. When (3.12) models the micropore diffusion, then $\beta(t) \sim t^{-1/2}$ is close to $t = 0$. Some approximations [20, 26] use $\beta(t) = \beta_{exp}(t) := rexp(-rt)$ with a rate $r > 0$. Note also that the kinetic model (3.5) can be written as a mild generalization of (3.12), where up to terms associated with the initial values of Υ, we have $\Upsilon := \frac{\partial f(\chi)}{\partial t} * \beta$ with $\beta = \beta_{exp}$.

Combine now (2.5) and (3.12) written together as

$$\frac{\partial \chi}{\partial t} + \frac{\partial \chi}{\partial t} * \beta + \mathbf{Adv} + \mathbf{Diff} = 0. \tag{3.13}$$

We recognize (3.13) as a double-porosity model [12, 59] for slightly compressible flow in oil and gas reservoirs with fractures and fissures. Also, see [99] for other multiscale analyses leading to a system part of which is similar to (3.13), and see [41, 95, 100, 105, 106, 147] for other computational models of double porosity.

To properly approximate the solutions to (3.13) we need to discretize the term $\frac{\partial \chi}{\partial t} * \beta$; here the difficulty is the singularity of β at $t = 0$. An appropriate discretization based on product integration rules was proposed, and the analysis of the resulting scheme for $\mathbf{Adv} \equiv 0$ carried out, in [105, 107].

However, in ECBM, we have $\mathbf{Diff} \equiv 0$, $\mathbf{Adv} \neq 0$. Now the appropriate numerical approximation of (3.13) falls in the class of schemes for scalar conservation laws with memory. In our recent work in [94] we developed convergence analysis for a scheme combining the Godunov scheme with the approximation of memory terms similar to that in [105, 107]. These results relate to the analysis of problems with memory in [30, 52], in which the smoothing effects of the memory terms on the solution to (3.13) are discussed. These, in turn, can be interpreted in ECBM from the physical standpoint as follows: the presence of the subscale diffusion into and out of the coal matrix has the potential to smooth out any sharp fronts, should they arise in the cleat.

As concerns multicomponent transport with memory, practical implementation results for (3.13) were reported for ECBM in [123, 149]. More accurate models combining multiporosity with IAS adsorption can be developed; see Sect. 3.4.

3.3 Adsorption Hysteresis

Desorption is a mechanism reverse to adsorption. In equilibrium, both are described by the same isotherm (3.3). Adsorption hysteresis occurs in special nonequilibrium circumstances in which these processes are described by different isotherms; see Fig. 1. The many theories that explain adsorption hysteresis do so either by studying phenomena in a single pore in the so-called independent pore theories, or by attributing the hysteresis to the presence of a complex interconnected pore network [13]. The first class of theories studies metastable states such as superheating and undercooling of a fluid undergoing phase transition in a single pore as well as the lack of symmetry of the gas–liquid interface upon filling and emptying. The second mechanism studies adsorption in pore networks and the pore-blocking and obstruction of desorption by the liquid remaining in the narrow necks of the pores [44, 115].

As concerns continuum computational models of adsorption hysteresis, one can proceed in one of (at least) two ways. The first, as in [122], uses separate isotherms $f_j, j = 1, 2$, as in (3.4), each with its own parameters $\Upsilon_j^{\max}, b_j$. However, it is not clear from [122] how the intermediate scanning curves are created when the desorption

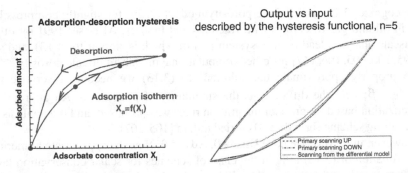

Fig. 1 *Left*: adsorption/desorption hysteresis similar to those in [55, 112, 122]. *Right*: a graph obtained with a Preisach model of hysteresis with $|A| = 5$ as described in [98]

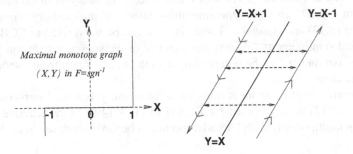

Fig. 2 Construction of an elementary hysteresis loop that serves as a building block for more complicated graphs. *Left*: a graph F. *Right*: hysteresis loop using F

occurs at an intermediate value of χ between the primary bounding curves; also, no analysis is available.

The second possibility is to consider a family of differential models of hysteresis [96, 98, 136] that is amenable to analysis. In the Preisach model of hysteresis, one generalizes (3.5) and considers a multivalued maximal monotone graph F instead of the function f in the form

$$\frac{\partial \Upsilon}{\partial t} + F(\Upsilon - \chi) \ni 0. \tag{3.14}$$

This construction allows an unambiguous representation of primary bounding and secondary scanning curves, and thanks to the convex–concave properties of the resulting hysteresis graph (see Fig. 1, right), it leads to well-posedness and higher regularity of solutions than those of usual conservation laws; see [98].

For terminology on monotone operators we refer to [125] and explain only briefly how (3.14) works to deliver, e.g., a graph such as in Fig. 1, right. Consider $F = sgn^{-1}$, with domain $D_F := [-1, 1]$; see Fig. 2. We can identify F with its graph and write $F := \{-1\} \times (-\infty, 0) \cup (-1, 1) \times \{0\} \cup \{1\} \times (0, \infty)$. The relation (3.14) means that $\Upsilon - \chi$ remains in the domain of F, and that the rate Υ_t is positive, negative, or zero, depending on whether $\Upsilon - \chi$ is equal to -1, $+1$, or is in $(-1, 1)$, respectively. Paraphrasing, given an increasing input χ, the output Υ is allowed either not to change or to increase along $\Upsilon = \chi - 1$. In contrast, for χ decreasing, Υ can only remain constant or decrease along $\Upsilon = \chi + 1$.

Now, using sgn^{-1} as a building block (but any other maximal monotone graph can be used), define $f_\alpha(s) := sgn^{-1}(\frac{2}{\alpha}s + 1)$, for some $\alpha > 0$, and let Υ_α satisfy an analogue of (3.14) in which F is replaced by f_α. This allows one to construct a complicated convex–concave graph via $\Upsilon = \sum_{\alpha \in A} \Upsilon_\alpha$, in which $A \subset \mathbb{R}_+$ is a finite collection of parameters; see the graph shown in Fig. 1, right, with $|A| = 5$. A more complicated, i.e., continuum, set A can be used instead, but this is irrelevant for a computational model. See [98] for details.

Consider a numerical scheme for (2.5) coupled with (3.14); the difficulty here is in understanding the differential inclusion in (3.14). This follows from the theory of evolution equations with monotone operators. We define the resolvent $J_\varepsilon := (I + \varepsilon F)^{-1}$ for every $\varepsilon > 0$ and the Yosida approximation $F_\varepsilon := \frac{1}{\varepsilon}(I - J_\varepsilon)$ to F. Both J_ε and F_ε are monotone Lipschitz functions with Lipschitz constants bounded by 1 and $\frac{1}{\varepsilon}$, respectively.

We propose two avenues toward the discretization of (3.14) beginning with $|A| = 1$. First, we can use a regularized model for (3.14), $\frac{\partial \Upsilon_\varepsilon}{\partial t} + F_\varepsilon(\Upsilon_\varepsilon - \eta) = 0$, in which we have replaced a multivalued F by the single-valued Yosida approximation F_ε, and an inclusion by an equation. Its numerical approximation should be implicit, because for $\varepsilon \to 0$, the ODE is very stiff. Either way, the error will depend on ε.

Second, we can consider an implicit in time discretization of (3.14),

$$\frac{\Upsilon_j^n - \Upsilon_j^{n-1}}{k} + F(\Upsilon_j^n - \chi_j^n) \ni 0. \tag{3.15}$$

This inclusion can be understood unambiguously using the resolvent $J_k = (I + kF)^{-1}$. Multiply (3.15) by k and subtract χ^n from both sides to see, after regrouping, that

$$\Upsilon_j^n - \chi_j^n + kF(\Upsilon_j^n - \chi_j^n) \ni \Upsilon_j^{n-1} - \chi_j^n.$$

Now apply the resolvent J_k to both sides to get

$$\Upsilon_j^n - \chi_j^n = J_k(\Upsilon_j^{n-1} - \chi_j^n). \tag{3.16}$$

This form, coupled with (3.7), is now amenable to analysis and implementation. The analysis reveals that the scheme is stable as long as $\lambda \leq 1$. However, questions remain as concerns the (order of) consistency of the scheme; see [93].

Now we describe how to solve (3.16). Substitute (3.16) into (3.7) to obtain a local nonlinear problem to be solved at each grid point j:

$$H(\chi_j^n) := \chi_j^n + \chi_j^n + J_k(\Upsilon_j^{n-1} - \chi_j^n) = A_j. \tag{3.17}$$

Since $2I + J_k$ is bijective, (3.17) is uniquely solvable. Since H is also piecewise linear and therefore *semismooth*, Newton's method will converge q-superlinearly; see [134, Propositions 2.12 and 2.25].

Now consider the case $|\mathcal{A}| > 1$. For each $\alpha \in \mathcal{A}$, we can choose a scheme based on Yosida approximation or on the resolvent. It appears most practical to use the latter scheme, since it does not depend on the regularization parameter. Thus, for each grid point j, n, a local system of equations composed of (3.7) and $|\mathcal{A}|$ variants of (3.16), with a sparse Jacobian, has to be solved.

Finally, adapting the construction above to arbitrary shape-scanning curves is part of our current work, since the symmetry of the graph F may be undesirable from the point of view of applications; compare the graphs in Fig. 1. Furthermore, multicomponent extensions are needed; see Sect. 3.4.

3.4 Multicomponent Transport with Adsorption

The models discussed in Sects. 3.2 and 3.3 are scalar, i.e., are defined for only one mobile component χ interacting with one adsorbed component Υ. In ECBM we have, however, several interacting components such as M, D, and N. For multiple components several issues arise: one concerning numerical approximation and analysis, and another concerning the data for (2.6).

We start with the first issue and discuss the multicomponent model as derived from (2.1). We assume for simplicity (3.1), (3.2) and ignore the presence of $C = N$, so that the nitrogen serves as a carrier only and does not get adsorbed. Further, let the flow displacement velocity $U_g \equiv const$ in the model be known. Thus we only consider equations for $C = D, M$. Further assume constant gas densities. Now, with rescaling and $\mathbf{Adv}_C = \nabla \cdot (U_g \chi_{gC})$ and $\mathbf{Diff}_C = \nabla \cdot (D_{gC} \nabla \chi_{gC})$, we can write the equations for $C = M, D$, each of which is similar to (2.5):

$$\frac{\partial}{\partial t}(\chi_{gM} + \chi_{aM}) + \mathbf{Adv}_M + \mathbf{Diff}_M = 0, \tag{3.18}$$

$$\frac{\partial}{\partial t}(\chi_{gD} + \chi_{aD}) + \mathbf{Adv}_D + \mathbf{Diff}_D = 0. \tag{3.19}$$

To complete this model, we need relationships between χ_{aC} and Υ_{aC} for $C = M, D$. These extend the equilibrium (3.3), kinetic (3.5), double porosity, or hysteretic relationships from Sects. 3.2, 3.3 and are coupled. For example, the equilibrium isotherm for $C = M$ depends on the other components $\Upsilon_M := \chi_{aM} = \chi_{aM}(\chi_{lM}, \chi_{lD})$ because the adsorption capacity of any porous system is finite, and thus the species compete for available adsorption sites.

The functional form of the multicomponent isotherms could be obtained experimentally. However, the number of necessary experiments appears impractical for systems with a large number of components. Instead, theoretical models for

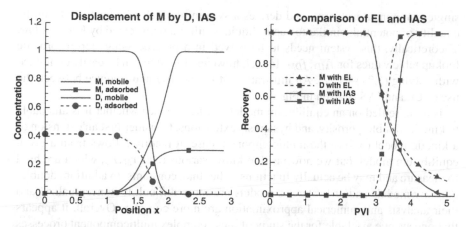

Fig. 3 *Left*: illustration of competitive adsorption when methane CH_4 ($C = M$) initially present in the coal matrix is displaced by the injected carbon dioxide CO_2 ($C = D$), which adsorbs more strongly. *Right*: comparison of CH_4/CO_2 recovery using extended Langmuir (EL) and ideal adsorbate theory (IAS). Plotted are breakthrough curves with respect to pore volume injected (PVI). Both examples use physical parameters from [62, 123], with nitrogen N_2 ignored

multicomponent mixtures have been proposed in the literature as extensions of single-component isotherms. For example, an extension of the Langmuir model is

$$\Upsilon_M = f_{MD}(\chi_M, \chi_D) := \Upsilon_M^{\max} \frac{b_M \chi_M}{1 + b_M \chi_M + b_D \chi_D}, \qquad (3.20)$$

with a similar equilibrium isotherm f_{DM} written for Υ_D, where we have used the simplified notation $\chi_C := \chi_{gC}; \Upsilon_C := \chi_{aC}$.

Assuming that the isotherms f_{MD}, f_{DM} are given, we consider briefly the numerical approximation of the equilibrium model. Here we can use an appropriate extension of the first-order Godunov scheme (3.7) for systems. Since the transport term itself is linear, we can simply use the upwind method, and the lack of an exact Riemann solver for the system is not an issue. However, we need to resolve the coupling in the storage terms, which, for **Diff** $\equiv 0$, extends (3.8) to a system of coupled algebraic equations

$$\chi_{M,j}^n + \Upsilon_{M,j}^n = \chi_{M,j}^n + f_{MD}(\chi_{M,j}^n, \chi_{D,j}^n) = A_{M,j}, \qquad (3.21)$$

$$\chi_{D,j}^n + \Upsilon_{D,j}^N = \chi_{D,j}^n + f_{DM}(\chi_{M,j}^n, \chi_{D,j}^n) = A_{D,j}. \qquad (3.22)$$

See an illustration of preferential adsorption and displacement of one component by the other in Fig. 3.

However, the extended Langmuir model (EL) (3.20) appears thermodynamically inconsistent [39]. A consistent more accurate model follows from the Ideal Adsorbate Solution (IAS) theory [39, 149]. IAS takes advantage of the easy-to-obtain single-component isotherms and derives a system of algebraic equations from the equality of potentials similar to vapor–liquid equilibria represented by Raoult's law. Theoretically, this system needs to be solved in a preprocessing step to provide lookup table values for f_{MD}, f_{DM}. In fact, however, this step can be easily combined with solving (3.21)–(3.22). An illustration of the nontrivial difference between the use of EL and IAS is given in Fig. 3.

Having settled on an equilibrium model, we need to see whether it is amenable to kinetic, double-porosity, and hysteresis extensions. Consider first an extension of a kinetic model (3.5) to the multicomponent case; it readily follows from a given equilibrium model, but we now need to know various rates r_M, r_D, which are hard to measure and may be actually functions rather than constants. In addition, double-porosity models and/or hysteresis models are more difficult to conceptualize, and their analysis and numerical approximation are more complex. Overall, it appears that one avenue available for the study of these complex multicomponent processes is to consider a completely different characterization of adsorption. See Sect. 5 for a discussion of adsorption at porescale.

4   Methane Hydrate Models

For methane hydrates [48, 76] one considers two mobile fluid phases l, g (brine, gas) and an immobile hydrate phase h (hydrate). In addition to the methane M, the main components of the fluids in the pore space are water W and salt S. Usually, the pore space is mainly saturated with water phase, with only a small amount of methane in the gas or hydrate phase being present.

The distribution of components between phases is governed by thermodynamics. Typically, we have

$$\chi_{gM} = 1 \text{ (gas phase contains methane only)},$$

$$\chi_{hM} + \chi_{hW} = 1 \text{ (both known for a fixed hydrate number)},$$

$$\chi_{lM} + \chi_{lW} + \chi_{lS} = 1 \text{ (unknown variables)}.$$

Phase behavior data for methane hydrates is available from various sources, although the models vary slightly in complexity and detail. In [101] we used phase behavior data from [48, 127, 131] for (i) the equilibrium (melting/dissociation) pressure $P^{EQ} = P^{EQ}(T, \chi_{lS})$ and (ii) maximum solubility $\chi_{lM}^{\max}(P, T, \chi_{lS})$ of methane M in liquid phase $p = l$; see Fig. 4 for an illustration. Other thermodynamic properties needed in a comprehensive model follow from an equation of state (EOS) [34, 42, 43, 48, 127, 131].

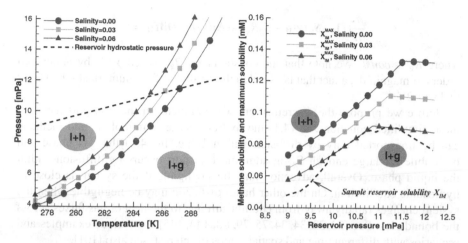

Fig. 4 *Left*: P–T diagram for hydrates used in [66,101]. *Right*: Maximum solubility χ_{lM}^{max} diagram for reservoir conditions

4.1 Simplified One-Equation Model

The simple model (2.7) explains well the main modeling constructs and solubility constraints. It can be obtained under the following simplifying assumptions. First, assume that P, T are given in a reservoir and that they follow a static distribution due to hydrostatic and geothermal gradients, respectively. Next, ignore the dependence of P^{EQ} and χ_{lM}^{max} on the salinity χ_{lS}, and set $\chi_{lS} \equiv const$. If pressure and temperature are low enough, i.e., above a certain depth below the Earth's surface, only a hydrate can form; see Fig. 4, left. On the other hand, below a certain critical depth, the pressure and temperature are higher, and only free gas can form. Finally, the hydrate and/or free gas can form only if the amount of methane N_M exceeds the amount associated with the maximum solubility at that depth; see Fig. 4, right.

In the hydrate zone we have $S_g = 0$ and $1 \geq S_l > 0$ with $1 > 1 - S_l = S_h \geq 0$. Assume constant $\phi, \rho_l, \rho_h, \chi_{hM}$. Equation (2.1), after rescaling, becomes then

$$\frac{\partial}{\partial t}((1 - S_h)\chi_{lM} + S_h R_h) + \mathbf{Adv}_M + \mathbf{Diff}_M = 0, \tag{4.1}$$

where $R_h \equiv const$, and where we have used $S_h + S_l = 1$. We now need a relationship between S_h and χ_{lM}, or their rates.

In equilibrium between the liquid and hydrate phases, we have the solubility constraint $\chi_{lM} \leq \chi_{lM}^{max}$ and the volume constraint $S_h \geq 0$. At most one of these inequalities can be sharp, which is expressed by $S_h(\chi_{lM}^{max} - \chi_{lM}) = 0$. This can be also written as $(\chi_{lM}, S_h) \in F := [0, \chi_{lM}^{max}) \times \{0\} \cup \{\chi_{lM}^{max}\} \times (0, 1)$; see Fig. 5.

A similar model can be formulated in the free gas zone for liquid–gas phase equilibria where $S_h = 0, S_l > 0, S_g \geq 0$. Such a model in which free gas phase may appear or disappear resembles the gas component part of the black-oil model [77, 97], carbon sequestration models such as [111], and hydrogen storage [50]:

$$\frac{\partial}{\partial t}((1 - S_g)\chi_{IM} + S_g R_g) + \mathbf{Adv}_M + \mathbf{Diff}_M = 0, \tag{4.2}$$

where $R_g \equiv const$. We note that the constants R_h, R_g exceed $\chi_{lm}^{\max}$ by about two orders of magnitude, a fact that is useful in the analysis of the numerical scheme for (4.1), (4.2).

Before we propose the discretization, we comment on the transport terms in these simplified equations. In (4.1) the hydrate phase is immobile, and methane can be transported only by diffusion within brine. In (4.2) the gas phase can be mobile for large enough S_g; in addition, transport occurs by diffusion within the liquid phase. Overall, advection may be negligible if the system is close to hydrostatic equilibrium. On the other hand, diffusion may be negligible when the system changes rapidly away from equilibrium due to the heat or gas fluxes across the boundaries. See [33, 33, 34, 34, 75, 79, 88, 144, 145] for various examples and scenarios with different time and spatial scales of interest; see also [101].

4.2 Numerical Model for the Simplified Case

Now we discuss the numerical approximation of the phase equilibria in the generic model (2.7), which includes (4.1) and (4.2) as special cases with $\Upsilon := S_h, R = R_h$, and $\Upsilon := S_g, R = R_g$, respectively, and with $\chi := \chi_{IM}$.

We discretize (2.7) in time and space and get

$$(1 - \Upsilon_j^n)\chi_j^n + R\Upsilon_j^n + k\mathbf{Diff}_j^{\bar{n}} = -k\mathbf{Adv}_j^{n-1} + (1 - \Upsilon_j^{n-1})\chi_j^{n-1} + R\Upsilon_j^{n-1},$$
$$(\chi_j^n, \Upsilon_j^n) \in F. \tag{4.3}$$

The usual treatment is to handle the advection explicitly and diffusion implicitly, so that $\bar{n} = n$, but in the exposition below we may use $\bar{n} = n - 1$. The computational realization of (4.3) is the crux of the algorithm.

Following the ideas in [50], we express (4.3) by recognizing that the mirror image of F, see Fig. 5, is a level curve of the min function, so that (4.3) is equivalent to

$$\min(\chi^{\max} - \chi_j^n, \Upsilon_j^n) = 0. \tag{4.4}$$

This characterization of the graph F now lends itself to a nonlinear iteration that can be applied to (4.3) and (4.4). Here we follow the ideas from [50] and consider a class of semismooth Newton methods described recently in [134]. For relevant background on complementarity conditions, see also [32, 60]. We outline this method below.

Let $\bar{n} = n - 1$, so we can write out the system (4.3) and (4.4) that has to be resolved at each grid point j, n as a local 2×2 system of equations

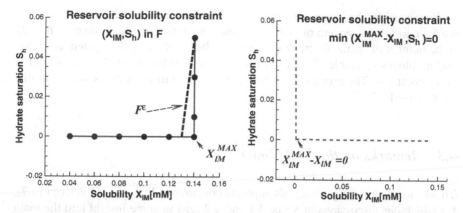

Fig. 5 *Left*: representation of a phase constraint between $\chi_{IM}^{\max}$ and S_h using the graph F, also its regularization F^ε, which was used in a kinetic model in [66]. *Right*: the mirror image of F and representation of the same constraint using the min function as in (4.4)

$$H(\chi_j^n, \Upsilon_j^n) = 0 \equiv \begin{cases} (1 - \Upsilon_j^n)\chi_j^n + R\Upsilon_j^n = A_j, \\ \min(\chi^{\max} - \chi_j^n, \Upsilon_j^n) = 0, \end{cases} \tag{4.5}$$

where A_j is given from (4.3).

Now the problem (4.5) can be solved by Newton's iteration. However, we observe the nondifferentiability of the second component of H across $\mathbb{R}^2 \supset A := \{(\chi, \Upsilon) : \Upsilon = \chi^{\max} - \chi\}$. Away from that singularity, H is smooth in both variables, and thus is a *semismooth function* on $B := \{(\chi, \Upsilon) : 0 \le \chi \le \chi^{\max}, 0 \le \Upsilon < 1\}$; see [134, Proposition 2.25].

We can calculate the Jacobian H' where it exists:

$$H' := \begin{cases} M_1 := \begin{bmatrix} 1 - \Upsilon & R - \chi \\ -1 & 0 \end{bmatrix}, \chi^{\max} - \chi > 0, \\[3mm] M_2 := \begin{bmatrix} 1 - \Upsilon & R - \chi \\ 0 & 1 \end{bmatrix}, \Upsilon > 0. \end{cases} \tag{4.6}$$

From (4.6) we can further compute its Bouligand subdifferential as the set $\partial_B H := \{M_1, M_2\}$ and its convex hull, the Clarke's generalized Jacobian ∂H,

$$\partial H := \left\{ \begin{bmatrix} 1 - \Upsilon & R - \chi \\ \alpha & \alpha + 1 \end{bmatrix}, -1 \le \alpha \le 0 \right\}.$$

We find then that any selection M from ∂H, thanks to $R \gg \chi_j^n$ and $\Upsilon_j^n < 1$, is always positive definite and thus nonsingular. Thus, from [134, Proposition 2.12], we conclude that Newton's method converges q-superlinearly.

Now consider the implicit treatment of diffusion, i.e., $\bar{n} = n$. At each time step n we have to solve a system of nonlinear equations with $2 \times M$ unknowns, where M is the number of spatial discretization points. The Jacobian of that system is sparse, and its blocks resemble H' in (4.5) because the constraints (4.4) have to hold at every point j, n. The properties of that system follow from analyses similar to that for $\bar{n} = n - 1$.

4.3 Remarks on the Full Model

Thanks to several simplifying assumptions concerning P, T, χ_{ls}, we derived (2.7), for which our discussions in Sects. 4.1 and 4.2 lead to some insight into the main difficulties of numerical hydrate modeling. These model simplifications must be removed for realistic simulations, and (2.7) must be complemented by an energy equation including the latent heat of phase transition. Furthermore, gas, hydrate, and brine phase compressibilities, as well as the dependence of transport properties on the primary variables, need to be accounted for. In addition, gas phase mobility and fluxes, also through the hydrate zone, cannot be ignored; in fact, capillary pressure needs to be included in the model, since from the observations of hydrates in subsea sediments it is known that methane bubbles percolate the hydrate zone through various free gas conduits.

The comprehensive models described in [29, 48, 76, 80, 87, 101, 132] are quite general and account for most of these dynamic effects using available experimental data and, when the experimental data is missing, using some heuristic relationships. For example, a model of hydrate evolution should account in some way for hydrate growth in pore space, and most models [48, 76] do so via heuristic relationships such as $\phi = \phi(S_h)$, $K = K(S_h)$, which are unsupported by experiments or first-principles models. The full mesoscale models for hydrate evolution are therefore lacking precision due to the lack of qualitative and quantitative information on the dynamics of hydrate formation and dissociation and its effects on porescale. See Sect. 5 for a discussion of qualitative and quantitative model elements following from porescale studies.

5 PoreScale Modeling

In Sects. 2–4 we discussed the need for accurate models of the dynamics of certain phenomena in ECBM and MH that are unavailable or hard to obtain experimentally. In ECBM models it is important to model porescale changes to the coal matrix under adsorption due to swelling [90, 91]. However, there is virtually no experimental work on porescale; this can be partially explained by the softness of coal and the difficulty in obtaining rock coal samples other than those made from pulverized coal. For MH, we are interested in phase transitions associated with hydrate formation and dissociation, a phenomenon not yet entirely understood; see [127, Chapters 3–5]. It is known that it proceeds in several steps, one of which includes the (Langmuir) adsorption of gas particles into the water structure surrounding the trapped gas molecules. The hydrate formation process includes heat and mass transfer and metastable states, while dissociation is an endothermic process because heat must be supplied to break the bonds between guest and water molecules and the hydrogen bonds between water molecules. Finally, there is substantial permeability decrease due to the presence of hydrate [29, 76, 88]; there are ongoing porescale imaging efforts [3, 117] showing that the models depend on the type of sediment and processes. However, experiments are difficult due to the instability of hydrates in standard conditions. In related work, [61] discusses the dependence between capillary invasion and fracture opening parameterized by grain size; however, changes in the geometry due to the transport of dissolved methane and/or hydrate formation are not accounted for. We argue that the nonlinear relationships necessary for ECBM and MH can be obtained from porescale computational models that can provide a virtual laboratory for much of the detailed data.

Until about a decade ago, most modeling in porous media relied on continuum models at mesoscale, carrying the quantitative information upward to reservoir scale by upscaling or multiscale modeling. There exist now various pore network models [8, 18, 73, 74], lattice–Boltzmann methods [121, 128, 139, 141], as well as continuum-based models [102, 103, 108] that simulate processes at porescale and supply data for the mesoscale models, or support them qualitatively (our references here are rather incomplete due to lack of space). Other discrete methods based on so-called first principles have been successful in various disciplines, and these include molecular dynamics (MD), density functional theory (DFT), and various Monte Carlo techniques; see [9, 54, 69, 113, 146]. More generally, statistical mechanics offers an approach in which "deterministic equations describing [large systems of] particles are replaced by assumptions on their statistical behavior" [19]. Particularly relevant for ECBM and MH are equilibrium lattice gas models, and specifically, those based on mean field theory (MFEQ) [22, 92, 130]. Below, we outline possible directions for ECBM and MH tied to some of our current work on porescale.

5.1 PoreScale Models of ϕ, K

Let the pore space $\omega = \omega_R \cup \omega_F = \bigcup_{i=1}^{N} \omega_i$ be a collection of rectangular cells (sites) ω_i, with random variables $\mathbf{t}, \mathbf{n}$ denoting matrix (rock) and fluid occupation variables; see Fig. 6 for an illustration. The geometry of ω either comes directly from tomography and x-ray analyses or is constructed synthetically based on some heuristics or on experimental structure factors such as mean, percolation, and two-point correlations. Note that $\mathbf{t}$ is usually called a *(quenched) disorder*. At a site i,

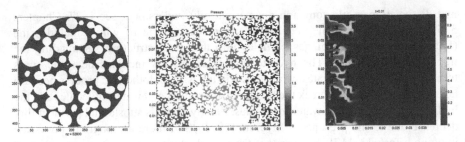

Fig. 6 *Left*: example of disorders t from porescale data from [109], courtesy of Dorthe Wilden-schild (OSU). *Middle* and *Right*: flow and transport simulations at porescale, in collaboration with Anna Trykozko (ICM UW)

$$t_i = \begin{cases} 1, \text{ cell is open to fluid,} \\ 0, \text{ cell is blocked by solid,} \end{cases} \tag{5.1}$$

$$n_i = \begin{cases} 1, \text{ cell is occupied by fluid,} \\ 0, \text{ cell is not occupied by fluid.} \end{cases} \tag{5.2}$$

We have $\omega_F := \bigcup_{i:t_i=1} \omega_i$, and we will define $|\omega_F| := \sum_i t_i$, with the porosity given by $\phi = \frac{|\omega_F|}{|\omega_F|+|\omega_R|} = \frac{N_F}{N}$. Also, the rock–fluid interface is denoted by $\gamma := \partial \omega_F \cap \partial \omega_R$. In discrete models, only t_i, n_i are important; continuum models are posed with respect to a position variable $y \in \omega_F$.

Let ϕ and K respond dynamically to the formation stress and to the presence of hydrates, adsorption in the matrix, and other phenomena, which we write as

$$K = g(\phi; \xi), \tag{5.3}$$

where ξ is a model-dependent collection of variables.

Now let ϕ_0, K_0 be some reference values of ϕ, K; assume for simplicity that K is isotropic. In ECBM the formula $\frac{K}{K_0} = (\frac{\phi}{\phi_0})^3$ for (5.3) or its generalizations have been used in [83, 91], but these do not account for the coal matrix swelling. For MH, $\phi = \phi_0(1 - S_h)$ shows the porosity changes due to the hydrate's presence via its saturation S_h, and $K = K_0(1 - S_h)^2$ for the *pore coating* models, or, via a more rapidly decaying relationship, $K = K_0(1 - S_h^2 + \frac{2(1-S_h)^2}{\log(S_h)})$ for the *pore filling* models [29, 76, 88]. These expressions are based on conceptual models of porous media as a bundle of capillaries, i.e., the Carman–Kozeny [17, 67] formula $K = \phi \frac{d_g^2}{12}$, where d_g is the grain size. These models and correlations, while valuable in simple circumstances, lack the precision for complex multicomponent phenomena.

A computational model to derive (5.3) begins with a given porescale geometry ω and a given ξ, and solves fluid-flow equations whose solutions we can then average to obtain (5.3). For an evolving ξ we need to solve as well the diffusive and advective transport models governing ξ at porescale; see an illustration in Fig. 6.

In particular, for ECBM and MH these need to account for adsorption and phase transition, respectively. A prevailing number of computational models of adsorption and phase transitions at porescale are discrete. However, some models [72, 116] allow the treatment of adsorption as a surface reaction modeled by a boundary condition of Robin or Neumann type posed on γ. These have proven to be most compatible in our studies thus far, while the discrete models appear to be better for modeling the dynamics of phenomena such as those in (2.6), discussed next.

5.2 Adsorption and Phase Transitions at PoreScale

The nonlinear relationships such as (3.3) or, more generally, (2.6) governing the dynamics of adsorption and phase transitions are usually known experimentally for some range of parameters and components. Precise quantitative models for any set of parameters and any number of components are needed for computational simulations, and ultimately for better understanding of the phenomena relevant for ECBM and MH at porescale and mesoscale. As mentioned above, we have had some preliminary success with the discrete rather than continuum models at porescale in this regard.

Specifically, the MFEQ theory accounts for the particles of fluid interacting with each other and adsorbing on the porous matrix and/or undergoing phase transitions. These interactions are expressed by a Hamiltonian, i.e., an energy functional with various product terms that are simplified in the mean field approximation. There are successful MFEQ models for adsorption and phase transitions developed in [64] based on the Ising model [68, 92]; they have been refined and compared to other methods in [53, 63, 85, 118–120, 142, 143]. For $N_R = N - N_F \gg 1$, there are several metastable equilibria in the Ising model [37, 92], and this leads to adsorption/desorption hysteresis and irreversibility of phase transition.

The MFEQ models for adsorption were derived and validated for single-component adsorbate in mesoporous glasses. We are working on extending them to multiple components in micropores and macropores in sediments and coals; these models will provide the porescale counterparts of the IAS–Langumir models from Sect. 3.4 and those with memory in Sect. 3.2. Another modification accounts for the swelling of the matrix, which is directly linked to the adsorption. Next, the discrete models of adsorption given by MFEQ provide scanning curves for adsorption/desorption hysteresis at arbitrary reversal points, and they have advantages over the models derived from experiments or the differential models described in Sect. 3.3, since they are easily extended to multiple components.

As concerns the phase transitions, the mean field approximation is capable of producing some good qualitative approximations to phase transitions and critical phenomena [92, 146]. We are working on a simple model of hydrate formation analogous to MFEQ models for adsorption. Compared to the classical Ising model of phase transition, it is defined in pore space, and its Hamiltonian can be formulated so as to promote either pore coating or filling.

5.3 Hybrid Models: Connection Between PoreScale and Mesoscale Models

The exposition above argues for implementation of porescale models to provide missing data for mesoscale simulations. One needs to consider several fundamental issues directly in determining their usefulness.

The first concerns the choice of porescale models. The discrete and stochastic models at porescale are fairly easy to implement, but are typically very computationally expensive, and their interpretation is difficult. On the other hand, continuum models at porescale and mesoscale require deep understanding in the startup phase, but the interpretation of the simulation results is fairly standard.

The second issue concerns the qualitative properties of their solutions. Once the porescale models are implemented, we should show, e.g., that the isotherms provided by MFEQ calculations have indeed the qualitative mathematical properties expected of the isotherms. An analogous issue concerns other results of porescale simulations.

The last but not the least issue is that of computational complexity. A common feature of the various porescale models is their large demand on computational time, which increases with $|\omega|$; the latter is as large as possible, since ω is intended to be a representative elementary volume (REV). In addition, we (may) need to consider several statistical realizations of the REVs. This combined computational complexity has precluded dynamic connections between porescale and mesoscale so far, while the mesoscale models are the only ones that can be used in optimization loops, reservoir characterization, parameter identification, and control.

We believe that it is possible for some of our porescale results to be encapsulated as library entries similar to lookup tables, with properties similar to the traditional heuristic models. As concerns the complexity, coarse-grained discrete models coupled to continuum models have been successful in other fields [40, 58, 84, 89]. Also, some porescale models can be perhaps implemented efficiently on modern computer architectures such as multicore or graphic processing units. However, there are as yet no universal ways to overcome the computational complexity of the hybrid models.

6 Summary and Acknowledgements

In this paper we have presented a collection of comprehensive and simplified models for ECBM and MH at mesoscale and underlined the main difficulties in their numerical approximation. Furthermore, we indicated the need for detailed dynamics information for some of the processes and proposed the use of porescale computations to derive these. We believe that some porescale models based on coarse graining, and a careful choice of REVs and their samples, can lead to dynamic hybrid models combining porescale and mesoscale models that can be very useful for the needs of MH and ECBM modeling.

The research for this paper has been partially supported by various sources and has benefited from discussions with several colleagues. We thank the IMA for hosting the author at workshops in May 2009 and October 2011 and as a short-term visitor in spring 2011. We also thank the Interdisciplinary Centre for Modeling at the University of Warsaw (ICM UW), where the author was a Fulbright Research Scholar in 2009–2010. Much of our knowledge of discrete models came from these visits. To Professor Knabner we owe the insight into semismooth Newton methods; we are also grateful to Professors Ben-Gharbia and Jaffre from INRIA for sharing more information on this topic. We also thank Professor Showalter (OSU) for many discussions on monotone operators over the years. Our understanding of continuum models of ECBM and MH benefited substantially from discussions with Professors Ceglarska-Stefańska and Czerw (University of Mining and Metallurgy, Cracow, Poland), and Professors Trèhu and Torres (COAS at OSU), respectively. Our collaborators Professors Wildenschild (OSU) and Trykozko (ICM UW) provided data for the images in Fig. 6. Research presented in this paper was partially supported by grants NSF 1115827 "Hybrid Modeling in Porous Media," NSF 0511190 "Model Adaptivity for Porous Media," and DOE 98089 "Modeling, Analysis, and Simulation of Preferential Flow."

References

1. IUPAC, International Union of Pure and Applied Chemistry http://www.iupac.org/.
2. Stomp: Subsurface Transport over Multiple Phases simulator website http://stomp.pnl.gov/, note = "[online; accessed 11-january-2010]".
3. The National Methane Hydrates R&D Program, http://www.netl.doe.gov/technologies/oil-gas/FutureSupply/MethaneHydrates/maincontent.htm,.
4. TOUGH Family of Codes: Availability and Licensing, http://esd.lbl.gov/TOUGH2/avail.html. [Online; accessed 11-January-2010].
5. Ocean Observatories Initiative, http://www.oceanleadership.org/programs-and-partnerships/ocean-observing/ooi/, 2007.
6. Sea expedition studies off the coast of northern Alaska, http://www.netl.doe.gov/publications/press/2009/09075-Beaufort_Sea_Expedition_Studies_Me.html, September 2009. NETL, U.S. Naval Research Laboratory and Royal Netherlands Institute for Sea Research, [Online; accessed 12-January-2010].
7. Neptune Canada, http://neptunecanada.ca/about-neptune-canada/neptune-canada-101.dot,, 2010.
8. David Adalsteinsson and Markus Hilpert. Accurate and efficient implementation of pore-morphology-based drainage modeling in two-dimensional porous media. *Transp. Porous Media*, 65(2):337–358, 2006.
9. M. P. Allen and D. J. Tildesley. *Computer Simulation of Liquids*. Oxford Science Publishers, 1987.
10. H. W. Alt and E. di Benedetto. Nonsteady flow of water and oil through inhomogeneous porous media. *Ann. Scuola Norm. Sup. Pisa Cl. Sci. (4)*, 12(3):335–392, 1985.
11. T. Arbogast. The existence of weak solutions to single porosity and simple dual-porosity models of two-phase incompressible flow. *Nonlinear Analysis, Theory, Methods and Applications*, 19:1009–1031, 1992.

12. Todd Arbogast, Jim Douglas, Jr., and Ulrich Hornung. Derivation of the double porosity model of single phase flow via homogenization theory. *SIAM J. Math. Anal.*, 21(4):823–836, 1990.

13. P. C. Ball and R. Evans. Temperature dependence of gas adsorption on a mesoporous solid: capillary criticality and hysteresis. *Langmuir*, 5(3):714–723, 1989.

14. J. W. Barrett, H. Kappmeier, and P. Knabner. Lagrange-Galerkin approximation for advection-dominated contaminant transport with nonlinear equilibrium or non-equilibrium adsorption. In *Modeling and computation in environmental sciences (Stuttgart, 1995)*, volume 59 of *Notes Numer. Fluid Mech.*, pages 36–48. Vieweg, Braunschweig, 1997.

15. John W. Barrett and Peter Knabner. Finite element approximation of the transport of reactive solutes in porous media. II. Error estimates for equilibrium adsorption processes. *SIAM J. Numer. Anal.*, 34(2):455–479, 1997.

16. John W. Barrett and Peter Knabner. An improved error bound for a Lagrange-Galerkin method for contaminant transport with non-Lipschitzian adsorption kinetics. *SIAM J. Numer. Anal.*, 35(5):1862–1882 (electronic), 1998.

17. Jacob Bear. *Dynamics of Fluids in Porous Media*. Dover, New York, 1972.

18. Martin Blunt. Flow in porous media pore-network models and multiphase flow. *Current Opinion in Colloid & Interface Science*, 6:197–207, 2001.

19. Oliver Bühler. *A brief introduction to classical, statistical, and quantum mechanics*, volume 13 of *Courant Lecture Notes in Mathematics*. New York University Courant Institute of Mathematical Sciences, New York, 2006.

20. Andreas Busch, Yves Gensterblum, Bernhard M. Krooss, and Ralf Littke. Methane and carbon dioxide adsorption-diffusion experiments on coal: upscaling and modeling. *International Journal of Coal Geology*, 60(2–4):151 – 168, 2004.

21. Andreas Busch, Yves Gensterblum, Bernhard M. Krooss, and Nikolai Siemons. Investigation of high-pressure selective adsorption/desorption behaviour of CO_2 and CH_4 on coals: An experimental study. *International Journal of Coal Geology*, 66(1–2):53 – 68, 2006.

22. Herbert B. Callen. *Thermodynamics, an introduction to physical theories of equilibrium thermostatics amd irreversible thermodynamics*. Wiley, 1960.

23. Grazyna Ceglarska-Stefanska and Katarzyna Zarebska. The competitive sorption of CO_2 and CH_4 with regard to the release of methane from coal. *Fuel Processing Technology*, 77–78:423 – 429, 2002.

24. Grazyna Ceglarska-Stefanska and Katarzyna Zarebska. Sorption of carbon dioxide-methane mixtures. *International Journal of Coal Geology*, 62(4):211 – 222, 2005.

25. Zhangxin Chen, Guan Qin, and Richard E. Ewing. Analysis of a compositional model for fluid flow in porous media. *SIAM J. Appl. Math.*, 60(3):747–777 (electronic), 2000.

26. C. R. Clarkson and R. M. Bustin. The effect of pore structure and gas pressure upon the transport properties of coal: a laboratory and modeling study. 1. isotherms and pore volume distributions. *Fuel*, 78(11):1333 – 1344, 1999.

27. C. R. Clarkson and R. M. Bustin. The effect of pore structure and gas pressure upon the transport properties of coal: a laboratory and modeling study. 2. adsorption rate modeling. *Fuel*, 78(11):1345 – 1362, 1999.

28. H. Class, R. Helmig, and P. Bastian. Numerical simulation of non-isothermal multiphase multicomponent processes in porous media 1. an efficient solution technique. *Advances in Water Resources*, 25:533–550, 2002.

29. M. B. Clennell, M. Hovland, J.S. Booth, P. Henry, and W. Winters. Formation of natural gas hydrates in marine sediments 1. conceptual model of gas hydrate growth conditioned by host sediment properties. *Journal of Geophysical Research*, 104:22,985–23,003, 1999.

30. B. Cockburn, G. Gripenberg, and S.-O. Londen. On convergence to entropy solutions of a single conservation law. *J. Differential Equations*, 128(1):206–251, 1996.

31. L.D. Connell and C. Detournay. Coupled flow and geomechanical processes during enhanced coal seam methane recovery through CO_2 sequestration. *International Journal of Coal Geology*, 77(1–2):222 – 233, 2009. CO_2 Sequestration in Coals and Enhanced Coalbed Methane Recovery.

32. Richard W. Cottle, Jong-Shi Pang, and Richard E. Stone. *The linear complementarity problem*. SIAM, 2009.
33. M. K. Davie and B. A. Buffett. A numerical model for the formation of gas hydrates below the seafloor. *Journal of Geophysical Research*, 106(B1):497–514, 2001.
34. M. K. Davie and B. A. Buffett. A steady state model for marine hydrate formation: Constraints on methane supply from pore water sulfate profiles. *Journal of Geophysical Research*, 108:B10, 2495, 2003.
35. C. N. Dawson. Godunov-mixed methods for advection-diffusion equations in multidimensions. *SIAM J. Numer. Anal.*, 30:1315–1332, 1993.
36. C. N. Dawson, C. J. van Duijn, and M. F. Wheeler. Characteristic-Galerkin methods for contaminant transport with nonequilibrium adsorption kinetics. *SIAM J. Numer. Anal.*, 31(4):982–999, 1994.
37. Pablo G. Debenedetti. *Metastable liquids. concepts and principles*. Princeton University Press, 1996.
38. G. R. Dickens. Rethinking the global carbon cycle with a large, dynamic and microbially mediated gas hydrate capacitor. *Earth Planet. Sci. Lett.*, 213, 2003.
39. Duong D. Do. *Adsorption analysis: equilibria and kinetics*. Imperial College Press, 1998.
40. Matthew Dobson and Mitchell Luskin. An optimal order error analysis of the one-dimensional quasicontinuum approximation. *SIAM J. Numer. Anal.*, 47(4):2455–2475, 2009.
41. J. Douglas, Jr., M. Peszyńska, and R. E. Showalter. Single phase flow in partially fissured media. *Transp. Porous Media*, 28:285–306, 1997.
42. Z. Duan. GEOCHEM_ORG website http://www.geochem-model.org. [Online; accessed 28-December-2009].
43. Z. Duan and S. Mao. A thermodynamic model for calculating methane solubility, density and gas phase composition of methane-bearing aqueous fluids from 273 to 523 k and from 1 to 2000 bar. *Geochimica et Cosmochimica Acta*, 70:3369–3386, 2006.
44. D. H. Everett. In D. H. Everett and F. S. Stone, editors, *The Structure and Properties of Porous Materials*. Butterworths, London, 1958.
45. R. W. Falta, K. Pruess, I. Javandel, and P. A Witherspoon. Numerical modeling of steam injection for the removal of nonaqueous phase liquids from the subsurface 1. numerical formulation. *Water Res. Research*, 28(2):433–449, 1992.
46. Masaji Fujioka, Shinji Yamaguchi, and Masao Nako. CO_2-ECBM field tests in the Ishikari coal basin of Japan. *International Journal of Coal Geology*, 82(3–4):287 – 298, 2010. Asia Pacific Coalbed Methane Symposium: Selected papers from the 2008 Brisbane symposium on coalbed methane and CO_2-enhanced coalbed methane.
47. T. S. Collett G. J. Moridis, S. R. Dallimore, T. Satoh, S. Hancock, and B. Weatherill. Numerical studies of gas production from several CH_4-hydrate zones at the Mallik site, Mackenzie Delta, Canada. Technical report.
48. S. K. Garg, J. W. Pritchett, A. Katoh, K. Baba, and T. Fijii. A mathematical model for the formation and dissociation of methane hydrates in the marine environment. *Journal of Geophysical Research*, 113:B08201, 2008.
49. Y. Gensterblum, P. van Hemert, P. Billemont, E. Battistutta, A. Busch, B. M. Kross, G. De Weireld, and K.-H. A. A. Wolf. European inter-laboratory comparison of high pressure CO_2 sorption isotherms ii: Natural coals. *International Journal of Coal Geology*, 84(2):115 – 124, 2010.
50. I. Ben Gharbia and J. Jaffre. Gas phase appearance and disappearance as a problem with complementarity constraints. Technical Report 7803, INRIA Research Report, November 2011.
51. A. L. Goodman, A. Busch, R. M. Bustin, L. Chikatamarla, S. Day, G. J. Duffy, J. E. Fitzgerald, K. A. M. Gasem, Y. Gensterblum, C. Hartman, C. Jing, B. M. Krooss, S. Mohammed, T. Pratt, R. L. Robinson Jr., V. Romanov, R. Sakurovs, K. Schroeder, and C. M. White. Inter-laboratory comparison ii: CO_2 isotherms measured on moisture-equilibrated Argonne premium coals at $55°C$ and up to 15 MPa. *International Journal of Coal Geology*, 72(3–4):153 – 164, 2007.

52. G. Gripenberg. Nonsmoothing in a single conservation law with memory. *Electron. J. Differential Equations*, pages No. 8, 8 pp. (electronic), 2001.
53. P. A. Monson H.-J. Woo, L. Sarkisov. Understanding adsorption hysteresis in porous glasses and other mesoporous materials. In *Characterization of porous solids VI ; Studies in surface science and catalysis*, volume 144. 2002.
54. J. M. Haile. *Molecular Dynamics Simulation*. Wiley, 1997.
55. Satya Harpalani, Basanta K. Prusty, and Pratik Dutta. Methane/CO_2 sorption modeling for coalbed methane production and CO_2 sequestration. *Energy & Fuels*, 20(4):1591–1599, 2006.
56. R. Helmig. *Multiphase flow and transport processes in the subsurface*. Springer, 1997.
57. P. Henry, M. Thomas, and M. B. Clennell. Formation of natural gas hydrates in marine sediments 2. Thermodynamic calculations of stability conditions in porous sediments. *Journal of Geophysical Research*, 104:23,005–23,022, 1999.
58. Desmond J. Higham. Modeling and simulating chemical reactions. *SIAM Rev.*, 50(2): 347–368, 2008.
59. Ulrich Hornung and Ralph E. Showalter. Diffusion models for fractured media. *J. Math. Anal. Appl.*, 147(1):69–80, 1990.
60. Kazufumi Ito and Karl Kunisch. *Lagrange multiplier approach to variational problems and applications*, volume 15 of *Advances in Design and Control*. Society for Industrial and Applied Mathematics (SIAM), Philadelphia, PA, 2008.
61. A. K. Jain and R. Juanes. Preferential mode of gas invasion in sediments: Grain scale mechanistic model of coupled multiphase fluid flow and sediment mechanics. *Journal of Geophysical Research*, 114:B08101, 2009.
62. Kristian Jessen, Wenjuan Lin, and Anthony R. Kovscek. Multicomponent sorption modeling in ECBM displacement calculations. *SPE 110258*, 2007.
63. E. Kierlik, P. A. Monson, M. L. Rosinberg, L. Sarkisov, and G. Tarjus. Capillary condensation in disordered porous materials: Hysteresis versus equilibrium behavior. *Phys. Rev. Lett.*, 87(5):055701, Jul 2001.
64. E. Kierlik, M. L. Rosinberg, G. Tarjus, and E. Pitard. Mean-spherical approximation for a lattice model of a fluid in a disordered matrix. *Molecular Physics: An International Journal at the Interface Between Chemistry and Physics*, 95:341–351, 1998.
65. G. R. King, T. Ertekin, and F. C. Schwerer. Numerical simulation of the transient behavior of coal-seam degasification wells. *SPE Formation Evaluation*, 2:165–183, April 1986.
66. Viviane Klein and Małgorzata Peszyńska. Adaptive multi-level modeling of coupled multiscale phenomena with applications to methane evolution in subsurface. In *Proceedings of CMWR XVIII in Barcelona, June 21–24, 2010*. available online at http://congress.cimne.com/CMWR2010/Proceedings, 2010. paper 47.
67. L. W. Lake. *Enhanced oil recovery*. Prentice Hall, 1989.
68. David Lancaster, Enzo Marinari, and Giorgio Parisi. Weighted mean-field theory for the random field Ising model. *J. Phys. A*, 28(14):3959–3973, 1995.
69. D. P. Landau and K. Binder. *A Guide to Monte-Carlo Simulations in Statistical Physics*. Cambridge, 2000.
70. R. J. Lenhard, M. Oostrom, and M. D. White. Modeling fluid flow and transport in variably saturated porous media with the STOMP simulator. 2. Verification and validation exercises. *Advances in Water Resources*, 18(6), 1995.
71. Randall J. LeVeque. *Numerical methods for conservation laws*. Lectures in Mathematics ETH Zürich. Birkhäuser Verlag, Basel, 1990.
72. J. Lewandowska, A. Szymkiewicz, K. Burzynski, and M. Vauclin. Modeling of unsaturated water flow in double-porosity soils by the homogenization approach. *Advances in Water Resources*, 27:283–296, 2004.
73. Li Li, Catherine A. Peters, and Michael A. Celia. Upscaling geochemical reaction rates using porescale network modeling. *Advances in Water Resources*, 29(9):1351 – 1370, 2006.
74. W. Brent Lindquist. Network flow model studies and 3D pore structure. In *Fluid flow and transport in porous media: mathematical and numerical treatment (South Hadley, MA, 2001)*, volume 295 of *Contemp. Math.*, pages 355–366. Amer. Math. Soc., Providence, RI, 2002.

75. X. Liu and P. B. Flemings. Passing gas through the hydrate stability zone at southern hydrate ridge, offshore oregon. *EPSL*, 241:211–226, 2006.
76. X. Liu and P. B. Flemings. Dynamic multiphase flow model of hydrate formation in marine sediments. *Journal of Geophysical Research*, 112:B03101, 2008.
77. Q. Lu, M. Peszyńska, and M. F. Wheeler. A parallel multi-block black-oil model in multi-model implementation. *SPE Journal*, 7(3):278–287, September 2002. SPE 79535.
78. Zofia Majewska, Grazyna Ceglarska-Stefanska, Stanislaw Majewski, and Jerzy Zietek. Binary gas sorption/desorption experiments on a bituminous coal: Simultaneous measurements on sorption kinetics, volumetric strain and acoustic emission. *International Journal of Coal Geology*, 77(1–2):90 – 102, 2009. CO_2 Sequestration in Coals and Enhanced Coalbed Methane Recovery.
79. A. Malinverno, M. Kastner, M. E. Torres, and U. G. Wortmann. Gas hydrate occurrence from pore water chlorinity and downhole logs in a transect across the northern cascadia margin (integrated ocean drilling program expedition 311. *Journal of Geophysical Research*, 113:B08103, 2008.
80. K. Masataka, N. Yukihiro, G. Shusaku, and A. Juichiro. Effect of the latent heat on the gas–hydrate/gas phase boundary depth due to faulting. *Bulletin of Earthquake Research Institute, University of Tokyo*, 73, 1998.
81. P. Massarotto, S. D. Golding, J.-S. Bae, R. Iyer, and V. Rudolph. Changes in reservoir properties from injection of supercritical CO_2 into coal seams – a laboratory study. *International Journal of Coal Geology*, 82(3–4):269 – 279, 2010. Asia Pacific Coalbed Methane Symposium: Selected papers from the 2008 Brisbane symposium on coalbed methane and CO_2-enhanced coalbed methane.
82. S. Mazumder, K. Wolf, P. van Hemert, and A. Busch. Laboratory experiments on environmental friendly means to improve coalbed methane production by carbon dioxide/flue gas injection. *Transport in Porous Media*, 75:63–92, 2008. 10.1007/s11242-008-9222-z.
83. Saikat Mazumder and Karl Heinz Wolf. Differential swelling and permeability change of coal in response to CO_2 injection for ECBM. *International Journal of Coal Geology*, 74(2):123– 38, 2008.
84. N. Moes, J. T. Oden, and K. Vemaganti. A two-scale strategy and a posteriori error estimation for modeling heterogeneous structures. In *On new advances in adaptive computational methods in mechanics*. Elsevier, 1998.
85. Peter Monson. Recent progress in molecular modeling of adsorption and hysteresis in mesoporous materials. *Adsorption*, 11:29–35(7), July 2005.
86. G. J. Moridis, T. Collett, S. R. Dallimore, T. Satoh, S. Hancock, and B. Weatherhill. Numerical studies of gas production from several ch_4-hydrate zones at the Mallik site, Mackenzie Delta, Canada. Technical report, LBNL-50257, 2002.
87. Kambiz Nazridoust and Goodarz Ahmadi. Computational modeling of methane hydrate dissociation in a sandstone core. *Chemical Engineering Science*, 62(22):6155 – 6177, 2007.
88. J. Nimblett and C. Ruppel. Permeability evolution during the formation of gas hydrates in marine sediments. *Journal of Geophysical Research*, 108:B9, 2420, 2003.
89. Christoph Ortner and Endre Süli. Analysis of a quasicontinuum method in one dimension. *M2AN Math. Model. Numer. Anal.*, 42(1):57–91, 2008.
90. Zhejun Pan and Luke D. Connell. A theoretical model for gas adsorption-induced coal swelling. *International Journal of Coal Geology*, 69(4):243 – 252, 2007.
91. Zhejun Pan, Luke D. Connell, and Michael Camilleri. Laboratory characterisation of coal reservoir permeability for primary and enhanced coalbed methane recovery. *International Journal of Coal Geology*, 82(3–4):252 – 261, 2010. Asia Pacific Coalbed Methane Symposium: Selected papers from the 2008 Brisbane symposium on coalbed methane and CO_2-enhanced coalbed methane.
92. Giorgio Parisi. *Statistical Field Theory*. Addison-Wesley, 1988.
93. M. Peszynska. Numerical model for adsorption hysteresis. manuscript.
94. M. Peszynska, Numerical scheme for a scalar conservation law with memory. *Numerical Methods for PDEs*, to appear.

95. M. Peszyńska. Finite element approximation of a model of nonisothermal flow through fissured media. In R. Stenberg M. Krizek, P. Neittaanmaki, editor, *Finite Element Methods*, Lecture Notes in Pure and Applied Mathematics, pages 357–366. Marcel Dekker, 1994.

96. M. Peszyńska. A differential model of adsorption hysteresis with applications to chromatography. In Jorge Guinez Angel Domingo Rueda, editor, *III Coloquio sobre Ecuaciones Diferenciales y Aplicaciones, May 1997*, volume II. Universidad del Zulia, 1998.

97. M. Peszyńska. The total compressibility condition and resolution of local nonlinearities in an implicit black-oil model with capillary effects. *Transport in Porous Media*, 63(1):201 – 222, April 2006.

98. M. Peszyńska and R. E. Showalter. A transport model with adsorption hysteresis. *Differential Integral Equations*, 11(2):327–340, 1998.

99. M. Peszyńska and R. E. Showalter. Multiscale elliptic-parabolic systems for flow and transport. *Electron. J. Diff. Equations*, 2007:No. 147, 30 pp. (electronic), 2007.

100. M. Peszynska, R. E. Showalter, and S.-Y. Yi. Homogenization of a pseudoparabolic system. *Applicable Analysis*, 88(9):1265–1282, 2009.

101. M. Peszyńska, M. Torres, and A. Tréhu. Adaptive modeling of methane hydrates. In *International Conference on Computational Science, ICCS 2010, Procedia Computer Science, available online via www.elsevier.com/locate/procedia and www.sciencdirect.com*, volume 1, pages 709–717, 2010.

102. M. Peszyńska and A. Trykozko. Convergence and stability in upscaling of flow with inertia from porescale to mesoscale. *International Journal for Multiscale Computational Engineering*, 9(2):215–229, 2011.

103. M. Peszyńska, A. Trykozko, and W. Sobieski. Forchheimer law in computational and experimental studies of flow through porous media at porescale and mesoscale. In *Mathematical Sciences and Applications*, volume 32 of *Current Advances in Nonlinear Analysis and Related Topics*, pages 463–482. GAKUTO Internat. Ser. Math. Sci. Appl., 2010.

104. M. Peszyńska, M. F. Wheeler, and I. Yotov. Mortar upscaling for multiphase flow in porous media. *Computational Geosciences*, 6:73–100, 2002.

105. Małgorzata Peszyńska. Analysis of an integro-differential equation arising from modelling of flows with fading memory through fissured media. *J. Partial Differential Equations*, 8(2):159–173, 1995.

106. Małgorzata Peszyńska. On a model of nonisothermal flow through fissured media. *Differential Integral Equations*, 8(6):1497–1516, 1995.

107. Małgorzata Peszyńska. Finite element approximation of diffusion equations with convolution terms. *Math. Comp.*, 65(215):1019–1037, 1996.

108. Malgorzata Peszynska, Anna Trykozko, and Kyle Augustson. Computational upscaling of inertia effects from porescale to mesoscale. In G. Allen, J. Nabrzyski, E. Seidel, D. van Albada, J. Dongarra, and P. Sloot, editors, *ICCS 2009 Proceedings, LNCS 5544, Part I*, pages 695–704, Berlin-Heidelberg, 2009. Springer-Verlag.

109. Mark L. Porter, Marcel G. Schaap, and Dorthe Wildenschild. Lattice-Boltzmann simulations of the capillary pressure-saturation-interfacial area relationship for porous media. *Advances in Water Resources*, 32(11):1632 – 1640, 2009.

110. K. Pruess. TOUGH2 – a general-purpose numerical simulator for multiphase fluid and heat flow. Technical Report LBL 29400, Lawrence Berkeley Laboratory, University of California, Berkeley, Calif, 1991.

111. Karsten Pruess and Julio Garcia. Multiphase flow dynamics during CO_2 disposal into saline aquifers. *Environmental Geology*, 42:282–295, 2002.

112. Basanta Kumar Prusty. Sorption of methane and CO_2 for enhanced coalbed methane recovery and carbon dioxide sequestration. *Journal of Natural Gas Chemistry*, 17(1):29 – 38, 2008.

113. D. C. Rapaport. *The art of molecular dynamics simulation*. Cambridge, 4 edition, 2009.

114. E. Ruckenstein, A. S. Vaidyanathan, and G. R. Youngquist. Sorption by solids with bidisperse pore structures. *Chemical Engrg. Science*, 26:1305–1318, 1971.

115. W. Rudzinski and D. H. Everett. *Adsorption of gases on heterogeneous surfaces*. Academic Press, 1992.

116. Emily M. Ryan, Alexandre M. Tartakovsky, and Cristina Amon. Investigating the accuracy of a Darcy scale model of competitive adsorption in a porous medium through SPH porescale modeling. In *Proceedings of CMWR XVIII in Barcelona, June 21–24, 2010*. available online at http://congress.cimne.com/CMWR2010/Proceedings, 2010. paper 94.
117. Marisa B. Rydzy, Mike L. Batzle, Keith C. Hester, Jim Stevens, and James J. Howard. Rock physics characterization of hydrate-bearing Ottawa sand f110. In *DHI/Fluid Consortium Meeting Fall 2010*, 2010.
118. L. Sarkisov and P. A. Monson. Computer simulations of phase equilibrium for a fluid confined in a disordered porous structure. *Phys. Rev. E*, 61(6):7231–7234, Jun 2000.
119. L. Sarkisov and P. A. Monson. Modeling of adsorption and desorption in pores of simple geometry using molecular dynamics. *Langmuir*, 17(24):7600–7604, 2001.
120. L. Sarkisov and P. A. Monson. Lattice model of adsorption in disordered porous materials: Mean-field density functional theory and Monte Carlo simulations. *Physical Review E (Statistical, Nonlinear, and Soft Matter Physics)*, 65(1):011202, 2002.
121. M. G. Schaap, M. L. Porter, B. S. B. Christensen, and D. Wildenschild. Comparison of pressure-saturation characteristics derived from computed tomography and lattice Boltzmann simulations. *Water Resour. Res.*, 43(W12S06), 2007.
122. C. J. Seto, G. T. Tang, K. Jessen, A. R. Kovscek, and F. M. Orr. Adsorption Hysteresis and its Effect on CO_2 Sequestration and Enhanced Coalbed Methane Recovery. *AGU Fall Meeting Abstracts*, pages D1542+, December 2006.
123. Ji-Quan Shi, Saikat Mazumder, Karl-Heinz Wolf, and Sevket Durucan. Competitive methane desorption by supercritical CO_2; injection in coal. *Transport in Porous Media*, 75:35–54, 2008. 10.1007/s11242-008-9209-9.
124. J. Q. Shi and S. Durucan. A bidisperse pore diffusion model for methane displacement desorption in coal by CO_2 injection. *Fuel*, 82:1219–1229, 2003.
125. R. E. Showalter. *Monotone operators in Banach space and nonlinear partial differential equations*, volume 49 of *Mathematical Surveys and Monographs*. American Mathematical Society, Providence, RI, 1997.
126. Hema J. Siriwardane, Raj K. Gondle, and Duane H. Smith. Shrinkage and swelling of coal induced by desorption and sorption of fluids: Theoretical model and interpretation of a field project. *International Journal of Coal Geology*, 77(1–2):188 – 202, 2009. CO_2 Sequestration in Coals and Enhanced Coalbed Methane Recovery.
127. E. D. Sloan and C. A. Koh. *Clathrate Hydrates of Natural Gases*. CRC Press, third edition, 2008.
128. Sauro Succi. *The Lattice Boltzmann equation for fluid dynamics and beyond*. Numerical Mathematics and Scientific Computation. The Clarendon Press Oxford University Press, New York, 2001. Oxford Science Publications.
129. C. E. Taylor, D. D. Link, and N. English. Methane hydrate research at NETL Research to make methane production from hydrates a reality. *JPSE*, 56:186–191, 2007.
130. Colin J. Thompson. *Classical Equilibrium Statistical Mechanics*. Oxford Science Publishers, 1988.
131. P. Tishchenko, C. Hensen, K. Wallmann, and C. S. Wong. Calculation of stability and solubility of methane hydrate in seawater. *Chemical Geology*, 219:37–52, 2005.
132. Ioannis N. Tsimpanogiannis and Peter C. Lichtner. Parametric study of methane hydrate dissociation in oceanic sediments driven by thermal stimulation. *Journal of Petroleum Science and Engineering*, 56(1–3):165 – 175, 2007. Natural Gas Hydrate / Clathrate.
133. Aslak Tveito and Ragnar Winther. On the rate of convergence to equilibrium for a system of conservation laws with a relaxation term. *SIAM J. Math. Anal.*, 28(1):136–161, 1997.
134. Michael Ulbrich. *Semismooth Newton methods for variational inequalities and constrained optimization problems in function spaces*, volume 11 of *MOS-SIAM Series on Optimization*. Society for Industrial and Applied Mathematics (SIAM), Philadelphia, PA, 2011.

135. Frank van Bergen, Pawel Krzystolik, Niels van Wageningen, Henk Pagnier, Bartlomiej Jura, Jacek Skiba, Pascal Winthaegen, and Zbigniew Kobiela. Production of gas from coal seams in the Upper Silesian Coal Basin in Poland in the post-injection period of an ECBM pilot site. *International Journal of Coal Geology*, 77(1–2):175 – 187, 2009. CO_2 Sequestration in Coals and Enhanced Coalbed Methane Recovery.

136. Augusto Visintin. *Differential models of hysteresis*, volume 111 of *Applied Mathematical Sciences*. Springer-Verlag, Berlin, 1994.

137. G. X. Wang, X. R. Wei, K. Wang, P. Massarotto, and V. Rudolph. Sorption-induced swelling/shrinkage and permeability of coal under stressed adsorption/desorption conditions. *International Journal of Coal Geology*, 83(1):46 – 54, 2010.

138. M. D. White, M. Oostrom, and R. J. Lenhard. Modeling fluid flow and transport in variably saturated porous media with the STOMP simulator. 1. Nonvolatile three-phase model description. *Advances in Water Resources*, 18(6), 1995.

139. D. Wildenschild, K. A. Culligan, and B. S. B. Christensen. Application of x-ray microtomography to environmental fluid flow problems. In U. Bonse, editor, *Developments in X-Ray Tomography IV*, volume 5535 of *Proc. of SPIE*, pages 432–441. SPIE, Bellingham, WA, 2004.

140. Karl-Heinz A. A. Wolf, Frank van Bergen, Rudy Ephraim, and Henk Pagnier. Determination of the cleat angle distribution of the RECOPOL coal seams, using CT-scans and image analysis on drilling cuttings and coal blocks. *International Journal of Coal Geology*, 73(3–4):259–272, 2008.

141. Dieter A. Wolf-Gladrow. *Lattice-gas cellular automata and lattice Boltzmann models*. Lecture Notes in Mathematics 1725. Springer, 2000.

142. Hyung-June Woo and P. A. Monson. Phase behavior and dynamics of fluids in mesoporous glasses. *Phys. Rev. E*, 67(4):041207, Apr 2003.

143. Hyung-June Woo, L. Sarkisov, and P. A. Monson. Mean-field theory of fluid adsorption in a porous glass. *Langmuir*, 17(24):7472–7475, 2001.

144. W. Xu. Modeling dynamic marine gas hydrate systems. *American Mineralogist*, 89: 1271–1279, 2004.

145. W. Xu and C. Ruppel. Predicting the occurrence, distribution, and evolution of methane hydrate in porous marine sediments. *Journal of Geophysical Research*, 104:5081–5095, 1999.

146. J. M. Yeomans. *Statistical Mechanics of Phase Transitions*. Oxford, 1992.

147. Son-Young Yi, Małgorzata Peszyńska, and Ralph Showalter. Numerical upscaled model of transport with non-separated scales. In *Proceedings of CMWR XVIII in Barcelona, June 21–24, 2010*. available online at http://congress.cimne.com/CMWR2010/Proceedings, 2010. paper 188.

148. Hongguan Yu, Guangzhu Zhou, Weitang Fan, and Jianping Ye. Predicted CO_2 enhanced coalbed methane recovery and CO_2 sequestration in China. *International Journal of Coal Geology*, 71(2–3):345 – 357, 2007.

149. Hongguan Yu, Lili Zhou, Weijia Guo, Jiulong Cheng, and Qianting Hu. Predictions of the adsorption equilibrium of methane/carbon dioxide binary gas on coals using Langmuir and Ideal Adsorbed Solution theory under feed gas conditions. *International Journal of Coal Geology*, 73(2):115 – 129, 2008.

150. Katarzyna Zarebska and Grazyna Ceglarska-Stefanska. The change in effective stress associated with swelling during carbon dioxide sequestration on natural gas recovery. *International Journal of Coal Geology*, 74(3–4):167 – 174, 2008.

Fast Algorithms for Bayesian Inversion

Sivaram Ambikasaran, Arvind K. Saibaba, Eric F. Darve,
and Peter K. Kitanidis

Abstract In this article, we review a few fast algorithms for solving large-scale
stochastic inverse problems using Bayesian methods. After a brief introduction to
the Bayesian stochastic inverse methodology, we review the following computational techniques, to solve large scale problems: the fast Fourier transform, the fast
multipole method (classical and a black-box version), and finallym the hierarchical
matrix approach. We emphasize that this is mainly a survey paper presenting
a few fast algorithms applicable to large-scale Bayesian inversion techniques,
applicable to applications arising from geostatistics. The article is presented at
a level accessible to graduate students and computational engineers. Hence, we
mainly present the algorithmic ideas and theoretical results.

Keywords Bayesian stochastic inverse modeling • Large-scale problems •
Geostatistical estimation • Numerical linear algebra • Fast Fourier transform •
Ast multipole method • Hierarchical matrices

AMS(MOS) Subject Classifications: Primary 1234, 5678, 9101112.

S. Ambikasaran (✉) • A.K. Saibaba
Institute for Computational and Mathematical Engineering, Huang Engineering Center 053B,
475, Via Ortega, Stanford University, Stanford, CA 94305-4042, USA

E.F. Darve
Mechanical Engineering, Durand 209, 496, Lomita Mall, Stanford University, Stanford,
CA 94305-4042, USA

P.K. Kitanidis
Civil and Environmental Engineering, Yang & Yamazaki Environment & Energy Building - MC
4020, 473 Via Ortega, Stanford, CA 94305-4042, USA
e-mail: peterk@stanford.edu

C. Dawson and M. Gerritsen (eds.), *Computational Challenges in the Geosciences*, The IMA
Volumes in Mathematics and its Applications 156, DOI 10.1007/978-1-4614-7434-0_5,
© Springer Science+Business Media New York 2013

1 Introduction

Large-scale stochastic inverse modeling has been drawing substantial attention in
recent times, especially in the context of Bayesian and geostatistical methodologies.
Inverse modeling is an essential ingredient in earth sciences modeling, particularly
in subsurface imaging, due to the fact that direct measurements are expensive
and sometimes impossible to obtain. By *inverse problems*, we mean estimation
of a function or many functions from information obtained from measurements.
Typically, the number of unknowns, m, is much larger than the number of observations, n. Since measurements are insufficient to uniquely obtain a *solution*, it is
often desirable to obtain a statistical ensemble of possible solutions. Hence, one
must obtain:

1. A representative depiction (or *best estimate*) of the unknown function.
2. A measure of accuracy, often in the form of a *variance* or a *confidence interval*.
3. Many probable solutions, often referred to as *conditional realizations*.

These inverse problems can be solved using the Bayesian geostatistics approach,
which combines data obtained through measurements with a stochastic model of the
structure of the function to be estimated. A principal advantage of this methodology
is that not only does it give the "best" estimate, but it also allows us to quantify
the uncertainty through confidence intervals or through functions that are samples
from the posterior distribution (conditional realizations). The results are particularly
useful for risk analysis.

However, a major hurdle of this approach is that when the number of unknowns,
m, to be estimated is large, the method becomes computationally expensive. The
bottleneck arises due to the fact that computational storage and matrix operations
involving large dense covariance matrices scale as $\mathcal{O}(m^2)$. This makes the method
computationally time-consuming. In this article, we present in detail a few existing
fast algorithms that reduce the computational cost for operations involving these
large dense covariance matrices.

The article is organized as follows. The next section, Sect. 2, gives a quick introduction to Bayesian stochastic linear inversion. The subsequent sections present
different fast algorithms to solve such inverse problems. Specifically, Sect. 3
discusses the fast Fourier transform, Sect. 4.1 discusses the original fast multipole
method (FMM), and Sect. 4.2 discusses a black-box version of the FMM. Section 5
discusses the hierarchical matrix approach.

There are also other efficient algorithms such as the panel-clustering method [33]
and wavelet-based methods [7,10,44,45], which could also be used in the context of
linear inversion for geostatistics. However, our goal was to present the algorithms
we have worked on in the context of Bayesian inversion and keep the article of
reasonable length.

2   Bayesian Stochastic Linear Inversion

Inverse problems are typically ill-posed because they involve few measurements and a large number of unknowns and require solving underdetermined linear systems. A popular technique for solving such underdetermined inverse problems is the Bayesian geostatistical approach [16,37–39,41,42,47,49]. In this section, we review the Bayesian stochastic linear inversion approach, which is a general method [42] for solving underdetermined linear systems arising out of linear inverse problems.

2.1   Prior

Consider a function $s(\mathbf{x})$ to be estimated. Its structure is represented through the prior probability density function. The basic model of the function to be estimated is taken as

$$s(x) = \sum_{k=1}^{p} f_k(x)\beta_k + \varepsilon(x). \tag{2.1}$$

The first term is the prior mean, where $f_k(x)$ are known functions, typically polynomials, and β_k, $k \in \{1, 2, \ldots, p\}$, are unknown drift coefficients that need to be determined. The second term $\varepsilon(x)$ is a random function with zero mean and characterized through a covariance function. This model is a popular choice in geostatistics and in other data analysis approaches. For instance, the conventional zonation/regression approach can be obtained as follows. Let p be the number of zones with

$$f_k = \begin{cases} 1, & \text{if in zone } k, \\ 0, & \text{otherwise,} \end{cases} \tag{2.2}$$

and we set the covariance of ε to zero. In the regression approach, the variability is described through deterministic functions. In stochastic approaches, one describes some of the variability through the stochastic part, while still retaining flexibility in the use of the deterministic part.

After discretization (e.g., through application of finite difference and finite element models), $s(x)$ is represented by a vector $s \in \mathbb{R}^{m \times 1}$. The mean of s is given by

$$\mathbb{E}[s] = X\beta \quad \text{(notation: } \mathbb{E} \text{ is the probabilistic expectation)}, \tag{2.3}$$

where $X \in \mathbb{R}^{m \times p}$ is the drift matrix and $\beta \in \mathbb{R}^{p \times 1}$ are the p unknown drift coefficients. The covariance matrix of s is given by

$$\mathbb{E}\left[(s - X\beta)(s - X\beta)^T\right] = Q. \tag{2.4}$$

The entries of the covariance matrix are given by $Q_{ij} = K(x_i, x_j)$, where $K(\cdot, \cdot)$ is a generalized covariance function, which must be conditionally positive definite. For

a more detailed discussion on permissible covariance kernels, we refer the reader to
the following references: [8, 35, 36, 40, 55].

2.2 Measurement Equation

The observation/measurement is related to the unknowns by the linear relation

$$y = Hs + v, \tag{2.5}$$

where $H \in \mathbb{R}^{n \times m}$ is the *observation/measurement matrix* and $v \in \mathbb{R}^{n \times 1}$ is a Gaussian
random vector of observation/measurement error independent of s with zero mean
and covariance matrix R, i.e., $v \sim \mathcal{N}(0, R)$. Typically, the matrix R is a diagonal
matrix, since each measurement is independent of other measurements.

The form mentioned in Eq. (2.5) is encountered frequently in practice, because
many important inverse problems are linear "deconvolution problems." Furthermore, many nonlinear problems are solved through a succession of linearized
problems.

The prior statistics of y are obtained as shown below. The mean is given by

$$\mu_y = \mathbb{E}[Hs + v] = H\mathbb{E}[s] + \mathbb{E}[v] = HX\beta = \Phi\beta, \tag{2.6}$$

the covariance is given by

$$\Psi = \mathbb{E}\left[(H(s - X\beta) + v)(H(s - X\beta) + v)^T\right] \tag{2.7}$$

$$= HQH^T + R, \tag{2.8}$$

and the y to s cross-covariance is

$$C_{ys} = \mathbb{E}\left[(H(s - X\beta) + v)(s - X\beta)^T\right] = HQ. \tag{2.9}$$

2.3 Bayesian Analysis

We assume that the prior probability distribution function is usually modeled as a
Gaussian

$$p'(s|\beta) \propto \exp\left(-\frac{1}{2}(s - X\beta)^T Q^{-1}(s - X\beta)\right). \tag{2.10}$$

To express that β is unknown a priori, its prior probability density function is
modeled uniformly over all space, i.e., $p'(\beta) \propto 1$, and thus

$$p'(s,\beta) \propto \exp\left(-\frac{1}{2}(s-X\beta)^T Q^{-1}(s-X\beta)\right). \tag{2.11}$$

The likelihood function is then given by

$$p(y|s) \propto \exp\left(-\frac{1}{2}(y-Hs)^T R^{-1}(y-Hs)\right). \tag{2.12}$$

Thus, the posterior probability density function is

$$p''(s,\beta) \propto \exp\left(-\frac{1}{2}(s-X\beta)^T Q^{-1}(s-X\beta)\right) \tag{2.13}$$

$$\times \exp\left(-\frac{1}{2}(y-Hs)^T R^{-1}(y-Hs)\right), \tag{2.14}$$

which is again a Gaussian. The negative log of the posterior probability density function is

$$\mathcal{L} = -\ln\left(p''(s,\beta)\right) \tag{2.15}$$

$$= \frac{1}{2}(s-X\beta)^T Q^{-1}(s-X\beta) \tag{2.16}$$

$$+ \frac{1}{2}(y-Hs)^T R^{-1}(y-Hs) + \text{constant.} \tag{2.17}$$

The posterior mean values, indicated by $\hat{s}$ and $\hat{\beta}$, minimize $\mathcal{L}$. Setting the partial derivative of $\mathcal{L}$ with respect to s and β to zero, we get the following equations:

$$(\hat{s}-X\hat{\beta})^T Q^{-1} - (y-H\hat{s})^T R^{-1}H = 0, \tag{2.18}$$

$$-\left(\hat{s}-X\hat{\beta}\right)^T Q^{-1}X = 0. \tag{2.19}$$

To bring the solution to a computationally convenient form, we introduce a vector $\xi \in \mathbb{R}^{n\times 1}$ defined through

$$y-HX\hat{\beta} = \Psi\xi. \tag{2.20}$$

Then we have

$$\boxed{\hat{s} = X\hat{\beta} + QH^T\xi.} \tag{2.21}$$

Substituting the above into Eq. (2.19), we get

$$\xi^T HX = 0. \tag{2.22}$$

Combining Eqs. (2.20) and (2.22), we get that

$$\begin{bmatrix} \Psi & \Phi \\ \Phi^T & 0 \end{bmatrix} \begin{bmatrix} \xi \\ \hat{\beta} \end{bmatrix} = \begin{bmatrix} y \\ 0 \end{bmatrix}. \tag{2.23}$$

Thus, the solution is obtained by solving a system of $n + p$ linear equations. The key equations are Eqs. (2.21) and (2.23).

The posterior covariance matrix, V, given by Eq. (2.24),

$$V = Q - QH^T P_{yy} HQ - XP_{bb}X^T - XP_{yb}^T (QH^T)^T - QH^T P_{yb}X^T, \tag{2.24}$$

enables us to quantify the uncertainty in the solution. The matrices P_{yy}, P_{yb}, P_{bb} in Eq. (2.24) are obtained as

$$\begin{bmatrix} P_{yy} & P_{yb} \\ P_{yb}^T & P_{bb} \end{bmatrix} = \begin{bmatrix} \Psi & \Phi \\ \Phi^T & 0 \end{bmatrix}^{-1}. \tag{2.25}$$

The diagonal entries of the matrix V enable us to quantify the uncertainty, since each diagonal entry represents the variance of each of the unknowns.

2.4 Computational Cost

Our goal is to obtain the best estimate $\hat{s}$ and the diagonal entries of the matrix V efficiently. The diagonal entries of V give us an estimate of the confidence intervals and thereby enable us to quantify uncertainty. To do so, we first need to solve the linear system in Eq. (2.23) and then obtain $\hat{s}$ using Eq. (2.21). A conventional direct algorithm, in which the number of measurements, n, is much smaller than the number of unknowns, m, would proceed as shown in Algorithm 1.

Since we have $p \ll n \ll m$, the total computational cost is $\mathcal{O}(nm^2 + n^2m)$. The cost to solve Eq. (2.23) is independent of m. Hence, once the linear system is constructed, it can be solved by any conventional direct method such as Gaussian elimination, which costs $\mathcal{O}((n + p)^3)$. Since we are interested in the case in which the number of measurements is relatively small, i.e., $n \approx 200$, we are not interested in optimizing the solving phase, since the cost is independent of m. Hence, the bottleneck is the cost $\mathcal{O}(nm^2)$, which arises from the computation of the matrix–matrix product $Q_H = QH^T$. The storage cost is dominated by the dense matrix Q, which costs $\mathcal{O}(m^2)$ to store. Hence, our focus will be on efficiently storing Q and efficiently constructing $Q_H = QH^T$, since this is the bottleneck in the algorithm. Some typical values of p, n, and m encountered in practice are $p \sim 1$–3, $n \sim 100$–200 and $m \sim 10^6$–10^9. The bottleneck in Bayesian stochastic inversion, as discussed above, is in computing a dense matrix–vector product, i.e., sums of the form

---

Algorithm 1 Conventional direct algorithm to solve large-scale linear inversion problem where the measurement operator H is dense

---

1. Compute $Q_H = QH^T$. The computational cost of this step is $\mathcal{O}(nm^2)$.
2. Compute $\tilde{\Psi} = HQ_H$. The computational cost of this step is $\mathcal{O}(n^2m)$.
3. Compute $\Psi = \tilde{\Psi} + R$. The computational cost of this step is $\mathcal{O}(n)$.
4. Compute $\Phi = HX$. The computational cost of this step is $\mathcal{O}(nmp)$.
5. Solve equation (2.23) to get ξ and β. The computational cost of this step is $\mathcal{O}((n+p)^3)$.
6. Compute $\hat{s} = X\beta + Q_H\xi$. The computational cost of this step is $\mathcal{O}(mp + mn)$.
7. Compute all the diagonal entries of V using the equation below:

$$V(i,i) = Q(i,i) - \underbrace{Q_H(i,:)P_{yy}(Q_H(i,:))^T}_{\mathcal{O}(n^2)} - \underbrace{X(i,:)P_{bb}(X(i,:))^T}_{\mathcal{O}(p^2)} - \underbrace{2X(i,:)P_{yb}^T(Q_H(i,:))^T}_{\mathcal{O}(pn)} \quad (2.26)$$

The cost of this step is $\mathcal{O}(n^2m)$.

$$\phi_j = \sum_{k=1}^{m} K(x_j, x_k)q_k, \quad \forall j \in \{1, 2, \ldots, m\}. \tag{2.27}$$

Direct computation of such sums requires $\mathcal{O}(m^2)$ operations. The main focus of this article is to reduce this computationally expensive step. In the subsequent sections, we look at a few ways to accelerate this computation and reduce the computational cost to almost linear complexity, i.e., $\mathcal{O}(m\log^\alpha m)$, where $\alpha \in \{0, 1\}$.

3 Fast Fourier Transform

In this section, we look at the best-known fast summation technique, namely, the fast Fourier transform algorithm, which was introduced [15] by Cooley and Tukey in 1965. Nowak et al. [22, 48] make use of the fast Fourier transform to address stochastic Bayesian inversion on a regular grid. The fast Fourier transform takes advantage of the fact that for a regular grid, the covariance matrix has a Toeplitz or block-Toeplitz structure, and this structure can be exploited to construct fast matrix–vector products in $\mathcal{O}(m\log m)$. The algorithm is also applied to a sandbox problem with a fine grid in [43]. However, one of the drawbacks of the method is that its extension to other grids is nontrivial, and most often in applications, such as those using the finite element approach, we rely on nonuniform unstructured grids.

First, let us consider the 1D case. The extension to 2D and 3D is relatively straightforward, once the 1D case is understood. Consider a set of equally spaced points $\{x_i\}_{i=1}^n$, such that $x_{i+1} - x_i = h, i = 1, \ldots, n-1$. A Toeplitz matrix T is an $n \times n$ matrix with entries such that $T_{ij} = t_{i-j}$, i.e., a matrix of the form

$$T_n = \begin{bmatrix} t_0 & t_{-1} & t_{-2} & \cdots & t_{-(n-1)} \\ t_1 & t_0 & t_{-1} & & \\ t_2 & t_1 & t_0 & & \vdots \\ \vdots & & & \ddots & \\ t_{n-1} & & \cdots & & t_0 \end{bmatrix}. \tag{3.1}$$

In particular, for stationary covariance kernels $Q_{ij} = \kappa(x_i, x_j) = \kappa((i-j)h)$ and for translation-invariant covariance kernels $Q_{ij} = \kappa(|i-j|h)$, both result in Toeplitz matrices. One of the features that make Toeplitz matrices interesting and useful for applications is that several fast schemes can be constructed for such matrices for matrix–vector multiplication, inversion, etc. A special form of the Toeplitz matrix is the circulant matrix C, in which each row is a cyclic shift of the row above it. In other words, the entries of the circulant matrices are given by $C_{ij} = c_{(j-i) \bmod n}$:

$$C_n = \begin{bmatrix} c_0 & c_1 & c_2 & & \cdots & c_{n-1} \\ c_{n-1} & c_0 & c_1 & c_2 & & \vdots \\ & c_{n-1} & c_0 & c_1 & \ddots & \\ \vdots & & \ddots & \ddots & \ddots & c_2 \\ & & & & & c_1 \\ c_1 & \cdots & & & c_{n-1} & c_0 \end{bmatrix}. \tag{3.2}$$

It is well known that circulant matrices can be diagonalized by the Fourier matrix by the following relation:

$$F_n C_n F_n^* = \Lambda_n = \begin{bmatrix} \lambda_1 & & & \\ & \lambda_2 & & \\ & & \ddots & \\ & & & \lambda_n \end{bmatrix}, \tag{3.3}$$

where F is the orthonormal complex symmetric Fourier matrix with entries

$$(F_n)_{jk} = \frac{1}{\sqrt{n}} \exp(2\pi i j k), \qquad i = \sqrt{-1}, \tag{3.4}$$

and

$$\Lambda_n e = \sqrt{n} F_n(C_n e_1), \qquad e_1 = [1, 0, \ldots, 0]^T, \tag{3.5}$$

and e is a vector of ones. This implies that the eigenvalues of the circulant matrices can be obtained by performing one FFT product of the first column of the circulant matrix. Thus, the cost of finding the eigenvalues of a circulant matrix is $\mathcal{O}(n \log n)$.

Moreover, it must be noted that one needs to store only the first row of the circulant matrix, with cost $\mathcal{O}(n)$.

Similarly, multiplying a circulant matrix by an appropriate-sized vector can be computed in $\mathcal{O}(n \log n)$ by performing FFT or inverse FFT, using the following relation:

$$C_n x = F_n^* \Lambda_n F_n x = F_n^* \overbrace{\Lambda_n F_n x}. \tag{3.6}$$

The matrix–vector product involving Toeplitz matrices can be performed efficiently by first embedding T_n in a circulant matrix

$$C_{2n} = \begin{bmatrix} T_n & B_n \\ B_n & T_n \end{bmatrix}, \tag{3.7}$$

where the two blocks B_n are computed such that C_{2n} is circulant of size $2n \times 2n$. Then the product can be computed in the following fashion:

$$C_{2n} \begin{bmatrix} x \\ 0 \end{bmatrix} = \begin{bmatrix} T_n & B_n \\ B_n & T_n \end{bmatrix} \begin{bmatrix} x \\ 0 \end{bmatrix} = \begin{bmatrix} T_n x \\ B_n x \end{bmatrix}. \tag{3.8}$$

In other words, we pad the vector x with zeros, then multiply this vector by the circulant matrix C_{2n}, and then $T_n x$ is obtained by extracting the first n components of this vector. The full algorithm can be obtained from [23].

It is easy to see that covariance matrices corresponding to regular grids in 2D result in block Toeplitz matrices, with Toeplitz subblocks (BTTB). A BTTB can be embedded inside a block circulant matrix with circulant subblocks (BCCB). A BCCB matrix has the form

$$C_{n_1 n_2} = \begin{bmatrix} C_0 & C_{n_2-1} & \cdots & C_2 & C_1 \\ C_1 & C_0 & C_{n_2-1} & & C_2 \\ \vdots & C_1 & C_0 & \ddots & \vdots \\ C_{n_2-2} & & \ddots & \ddots & C_{n_2-1} \\ C_{n_2-1} & C_{n_2-2} & \cdots & C_1 & C_0 \end{bmatrix}, \tag{3.9}$$

where each C_i is an $n_1 \times n_1$ circulant matrix, and we set $n = n_1 n_2$. Then $C_{n_1 n_2}$ can be diagonalized by $F_{n_2} \otimes F_{n_1}$, which results in a block diagonal matrix whose diagonal blocks are diagonal matrices

$$(F_{n_2} \otimes F_{n_1}) C_{n_1 n_2} (F_{n_2} \otimes F_{n_1})^* = \begin{bmatrix} \sum_{j=0}^{n_2-1} \Lambda_j & & & \\ & \sum_{j=0}^{n_2-1} \omega^j \Lambda_j & & \\ & & \ddots & \\ & & & \sum_{j=0}^{n_2-1} \omega^{(n_2-1)j} \Lambda_j \end{bmatrix}, \quad (3.10)$$

where $\omega = \exp(2\pi i / n_1)$, $i = \sqrt{-1}$, and

$$F_{n_1} C_j F_{n_1}^* = \Lambda_j \qquad j = 0, \ldots, n_2 - 1.$$

Therefore, the eigenvalues of the BCCB matrix can be obtained by a 2D FFT, which costs $\mathcal{O}(n \log n)$. Similarly, matrix–vector products with BCCB can be performed in $\mathcal{O}(n \log n)$. Again, one needs to store only the first row of the BCCB matrix, which costs $\mathcal{O}(n)$.

With similar notation to that of a BCCB matrix, we define a BTTB matrix having the form

$$T_{n_1 n_2} = \begin{bmatrix} T_0 & T_{-1} & \ldots & T_{-n_2+2} & T_{-n_2+1} \\ T_1 & T_0 & T_{-1} & & T_{-n_2+2} \\ \vdots & T_1 & T_0 & \ddots & \vdots \\ T_{n_2-2} & & \ddots & \ddots & T_{-1} \\ T_{n_2-1} & T_{n_2-2} & \ldots & & T_0 \end{bmatrix}, \quad (3.11)$$

where each T_j is an $n_1 \times n_1$ block. To multiply $T_{n_1 n_2}$ by an appropriate-sized vector x, we have to perform a double circulant embedding: first we embed the block T_j in a circulant form to obtain a block Toeplitz matrix with circulant blocks, and then we embed the resulting block Toeplitz matrix in a block circulant matrix, which results in a BCCB matrix. Then after appropriately zero-padding the vector, we can compute the matrix–vector product $T_{n_1 n_2} x$ in $\mathcal{O}(n \log n)$ time by this embedding and extract the relevant components. Extension to three and higher dimensions is quite simple and will not be stated here.

4   Fast Multipole Method

The fast multiple method (FMM) is a computational technique to calculate matrix–vector products, or equivalently, sums of the form

$$f(x_i) = \sum_{j=1}^{N} K(x_i, y_j) \sigma_j, \quad (4.1)$$

where $i \in \{1,2,3,\ldots,M\}$, in $\mathcal{O}(M+N)$ operations as opposed to $\mathcal{O}(MN)$ with a controllable error ε. The FMM was originally introduced by Greengard et al. [26]. They developed this technique for the kernel $K(x,y) = \frac{1}{|x-y|}$ using Legendre polynomial expansions and spherical harmonics. The literature on FMM is vast, and we refer the reader to a few papers:[2, 3, 13, 14, 17–21, 27, 46, 58].

Initially, the FMM was used for the treatment of integral operators (boundary element method, or BEM) and the solution of the resulting linear systems, which arise in solving partial differential equations, such as the Laplace, Helmholtz, and Maxwell equations. It has been applied in a variety of fields including molecular dynamics, astrophysics, elastic materials, and graphics. In all these cases, the FMM is used, as part of an iterative solution of linear systems (GMRES, conjugate gradient), to accelerate matrix–vector products. This is often the part of the calculation that is computationally the most expensive. Recently, the FMM has also been applied in the context of large-scale stochastic inversion [1] for geostatistical application. In the case of Bayesian inversion, the FMM will be used to construct the linear system, i.e., more precisely in constructing QH^T, as discussed in the previous section.

4.1 Classical FMM

As mentioned above, the original FMM is based on certain analytic expansions of the kernel $K(r)$, for example with spherical harmonics, powers of r, Bessel functions. We will present the classical FMM for the two-dimensional case, since it is conceptually easy to understand. The ideas and methodology carry over to the three-dimensional case as well. The presentation of this subsection is based on [2]. We will consider the logarithmic kernel given by $K(x_j, x_k) = -\log(\|x_j - x_k\|)$. Also, we will use the language of electrostatics, since the original FMM was developed for electrostatic interactions. If a charge of strength q is located at $(x_0, y_0) \in \mathbb{R}^2$, then the potential at $(x,y) \in \mathbb{R}^2$ is given by

$$\phi_{q,z_0}(z) = -q\log(\|z - z_0\|) = \mathrm{Re}\left(-q\log(z - z_0)\right), \tag{4.2}$$

where $z = x + iy$, $z_0 = x_0 + iy_0$, and $z, z_0 \in \mathbb{C}$. Let $\psi_{q,z_0}(z) = q\log(z - z_0)$, i.e., $\phi_{q,z_0}(z) = -\mathrm{Re}\left(\psi_{q,z_0}(z)\right)$. Now note that for every z such that $|z| > |z_0|$, we have

$$\psi_{q,z_0}(z) = q\log(z - z_0) = q\left(\log(z) - \sum_{k=1}^{\infty} \frac{1}{k}\left(\frac{z_0}{z}\right)^k\right). \tag{4.3}$$

This analytic expansion is the basis for computing the multipole expansion [26, 52] when we have a collection of charges.

Lemma 4.1. *Consider a set of n charges of strength q_k located at z_k, where $k \in \{1, 2, \ldots, n\}$. Let $R = \max_k |z_k|$. For every z such that $|z| > R$, the potential $\phi(z)$ due to the n charges is given by*

$$\phi(z) = q \log(z) - \sum_{k=1}^{\infty} \frac{a_k}{k z^k}, \tag{4.4}$$

where $q = \sum_{j=1}^{n} q_j$ and $a_k = \sum_{j=1}^{n} q_j z_j^k$.

More importantly, for every $p \in \mathbb{Z}^+$, if we set

$$\phi_p(z) = q \log(z) - \sum_{k=1}^{p} \frac{a_k}{k z^k}, \tag{4.5}$$

then we have the following error bound, given by Eq. (4.6):

$$|\phi(z) - \phi_p(z)| \leq \frac{\sum_{j=1}^{n} |q_j|}{(p+1)\left(1 - \left|\frac{r}{z}\right|\right)} \left|\frac{r}{z}\right|^{p+1}. \tag{4.6}$$

Note that $\left|\frac{r}{z}\right| < 1$.

In the above, if we have $|z| > 2|r|$, we then get that

$$|\phi(z) - \phi_p(z)| \leq \frac{\sum_{j=1}^{n} |q_j|}{(p+1)2^p} \tag{4.7}$$

4.1.1 The *m log m* Algorithm

We will make use of the *multipole expansion* to describe the $\mathcal{O}(m \log m)$ algorithm first and then proceed to describing the FMM. The key ingredient of the $\mathcal{O}(m \log m)$ algorithm is the "divide and conquer" approach. This is done recursively taking advantage of the *multipole expansion* to attain a computational complexity of $\mathcal{O}(m \log m)$.

For pedagogical reasons, to convey the idea, we will consider that the charges are distributed more or less uniformly inside a square domain. The overall idea behind the $\mathcal{O}(m \log m)$ algorithm is to recursively divide the domain into clusters and make use of the *multipole expansion* for clusters that are *sufficiently far apart*. This will be made precise soon, but first we introduce some definitions and terminology that will be useful when we discuss the algorithm.

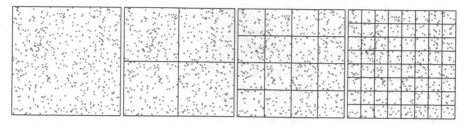

Fig. 1 A square domain with charges indicated by *red dots*. The *leftmost* figure is the partition of the domain at level 0, while the *rightmost* is the partition of the domain at level 3 (color figure online)

From Fig. 1, we see that there is a natural tree structure, whereby the cluster at level $\ell + 1$ is obtained by subdividing the cluster at level ℓ.

Definition 4.1. A square at any level is termed a *cluster*. For instance, in Fig. 1, the entire domain at level 0 is a cluster, each of the four squares at level 1 is a cluster, and so on. In general, there are 4^ℓ clusters at level ℓ.

Definition 4.2. Consider two clusters $\mathscr{C}_1$ and $\mathscr{C}_2$. If $\mathscr{C}_1$ is at level ℓ and if $\mathscr{C}_2$ is at level $\ell + 1$ and if $\mathscr{C}_2 \subseteq \mathscr{C}_1$, then $\mathscr{C}_1$ is termed the *parent* of $\mathscr{C}_2$, while $\mathscr{C}_2$ is termed the *child* of $\mathscr{C}_1$.

Definition 4.3. Two clusters are said to be *neighbors* of each other if the clusters are at the same level and share a boundary.

Definition 4.4. Two clusters are said to be *well separated* if the clusters are at the same level and are not neighbors.

Definition 4.5. A cluster $\mathscr{C}_1$ is in the *interaction list* of $\mathscr{C}_2$ if $\mathscr{C}_1$ is well separated from $\mathscr{C}_2$ and the parents of $\mathscr{C}_1$ and $\mathscr{C}_2$ are neighbors.

To start off, the entire square domain is our cluster. This is level 0 in the algorithm. At the next level, i.e., at level 1, the entire square domain is divided into four equal clusters, each a smaller square domain. In general, the clusters at level $l + 1$ are obtained by dividing each cluster at level l into four equal squares. This is pictorially described in Figs. 2 and 3. Note that for a cluster $\mathscr{C}$ at a given level, we can partition the remaining clusters at the same level into three mutually disjoint sets:

- $\mathscr{N}_{\mathscr{C}}$: the set containing clusters that are neighbors to $\mathscr{C}$;
- $\mathscr{I}_{\mathscr{C}}$: the set containing clusters that are in the interaction list of $\mathscr{C}$;
- $\mathscr{W}_{\mathscr{C}}$: the set containing the rest, i.e., the set containing clusters that are well separated from $\mathscr{C}$ but are not in the interaction list of $\mathscr{C}$.

The three sets for a cluster at level 3 are indicated in Fig. 3.

Once we have such a partition of the domain, we can compute the interaction between clusters that are not neighbors using the truncated multipole expansion (4.5).

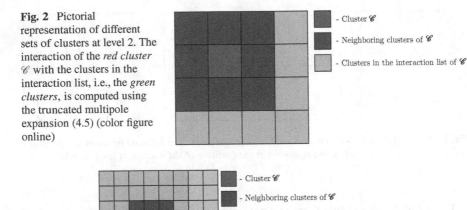

Fig. 2 Pictorial representation of different sets of clusters at level 2. The interaction of the *red cluster* $\mathscr{C}$ with the clusters in the interaction list, i.e., the *green clusters*, is computed using the truncated multipole expansion (4.5) (color figure online)

- Cluster $\mathscr{C}$
- Neighboring clusters of $\mathscr{C}$
- Clusters in the interaction list of $\mathscr{C}$

- Cluster $\mathscr{C}$
- Neighboring clusters of $\mathscr{C}$
- Clusters in the interaction list of $\mathscr{C}$
- Remaining clusters at the same level as $\mathscr{C}$

Fig. 3 Pictorial representation of different sets of clusters at level 3. The interaction of the *red cluster* $\mathscr{C}$ with the clusters in the interaction list, i.e., the *green clusters*, is computed using the truncated multipole expansion (4.5). The interaction of the *red cluster* with the *pink clusters* has already been computed at the previous level, level 2 (color figure online)

At levels 0 and 1, there are no well-separated clusters. From the second level onward, for every cluster there are well-separated clusters. Note that at this level (Fig. 2), all the well-separated clusters are in the interaction list as well. Hence, from the second level onward, we can use the truncated multipole expansion (4.5) to compute the interaction of the cluster with the clusters in its interaction list. Now we need to compute the interaction of the cluster with its neighbors. To do this, we subdivide each of the clusters at level 2 into four smaller clusters. Now at level 3, as in Fig. 3, note that the interaction between the red cluster and the pink clusters has already been computed at the previous level 2 (see Fig. 2), since the parent of the pink cluster was in the interaction list of the parent of the red cluster at level 2. Compute the interaction of the red cluster at level 3, as in Fig. 3, with the clusters in its interaction list, i.e., the green clusters in Fig. 3, using the truncated multipole expansion. Perform this operation recursively to compute the interaction of the red cluster with its neighbors.

Hence, at every level, form the truncated multipole expansion for each cluster and evaluate this for each charge in the interaction list of the cluster. The recursion proceeds till the number of charges in each cluster is approximately of size p. As stated earlier, since we take the distribution of charges to be more or less uniform, there will be $\mathcal{O}(\log_4(m/p))$ levels. At each level, the computational complexity to compute the interaction of all of the clusters with clusters in its interaction list is $\mathcal{O}(m)$. The complexity $\mathcal{O}(m)$ is due to the fact that the cost to compute the truncated multipole expansion is pm and each cluster has at most 27 other clusters in its

Fig. 4 An unbalanced tree
for a nonuniform distribution
of charges

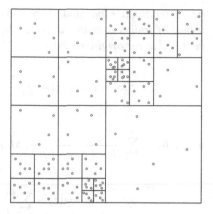

interaction list. Hence, the total computational cost at each level is $pm + 27pm$. At the last level, typically termed the leaf level, there are approximately m/p clusters, each containing roughly p charges. Each cluster has at most 8 neighbors, and hence at the leaf level there is an additional cost of $8pm$. Hence, the total computational complexity scales roughly as

$$28pm\log m + 8pm. \tag{4.8}$$

4.1.2 Unbalanced Tree Structure

Often in many practical applications, the distribution of charges is nonuniform. This would in turn imply that the partition of the space must be slightly different to take into account the nonuniform distribution of the charges. An example of such a partition is shown in Fig. 4. The partition of a cluster is terminated when the number of charges within the cluster reaches a threshold. The algorithm stated in the previous section carries forward for this unbalanced tree as well, though the complexity is not exactly $\mathcal{O}(pm\log m)$. However, on average, the computational complexity is still $\mathcal{O}(pm\log m)$.

4.1.3 The Fast Multipole Method

We described the $\mathcal{O}(m\log m)$ algorithm in Sect. 4.1.1. In this section, we describe the FMM, which further reduces the complexity to $\mathcal{O}(m)$. To reduce the complexity further to $\mathcal{O}(m)$, we need a few more results.

Lemma 4.2. *Let*

$$\phi(z) = a_0 \log(z - z_0) + \sum_{k=1}^{\infty} \frac{a_k}{(z - z_0)^k} \tag{4.9}$$

be the multipole expansion of the potential due to a set of m charges located inside a ball of radius R centered at z_0. Then for all z outside a ball of radius $R + |z_0|$ centered at the origin, we have

$$\phi(z) = a_0 \log(z) - \sum_{j=1}^{\infty} \frac{a_0}{j} \left(\frac{z_0}{z}\right)^j + \sum_{j=1}^{\infty} b_j (j-1)! \left(\frac{z_0}{z}\right)^j, \qquad (4.10)$$

where $b_j = \sum_{k=1}^{j} \dfrac{a_k}{(k-1)!(j-k)!z_0^k}$. More importantly, for $p \in \mathbb{Z}^+$, if we let

$$\phi_p(z) = a_0 \log(z) - \sum_{j=1}^{p} \frac{a_0}{j} \left(\frac{z_0}{z}\right)^j + \sum_{j=1}^{p} b_j (j-1)! \left(\frac{z_0}{z}\right)^j, \qquad (4.11)$$

then we have the following error bound:

$$\left|\phi_p(z) - \phi(z)\right| \leq \left(\frac{\displaystyle\sum_{k=1}^{m} |q_k|}{1 - \left|\dfrac{|z_0| + R}{z}\right|}\right) \left|\frac{|z_0| + R}{z}\right|^{p+1}. \qquad (4.12)$$

Lemma 4.2 gives us a way of translating the center of the multipole expansion. The following lemma gives the conversion of a multipole expansion into a local expansion.

Lemma 4.3. *Given n charges $q_1, q_2, \ldots, q_n$ located inside the circle D_1 with radius R and center at z_0, and that $|z_0| > (c+1)R$ with $c > 1$, then the multipole expansion in Eq. (4.10) converges inside the circle, D_2, of radius R centered at the origin, and the potential is given by the power series*

$$\phi(z) = \sum_{l=0}^{\infty} b_l z^l \qquad (4.13)$$

with

$$b_0 = a_0 \log(-z_0) + \sum_{k=1}^{\infty} \frac{a_k}{z_0^k} (-1)^k, \qquad (4.14)$$

and for $l \geq 1$,

$$b_l = -\frac{a_0}{l \cdot z_0^l} + \frac{1}{z_0^l} \sum_{k=1}^{\infty} \frac{a_k}{z_0^k} \frac{(l+k-1)!}{l!(k-1)!} (-1)^k. \qquad (4.15)$$

Further, if we let $\tilde{\phi}_p = \sum_{l=0}^{p} b_l z^l$, we then have the following error bound:

$$|\phi(z) - \phi_p(z)| < \frac{\sum_{k=1}^{n} |q_k| \left(4e(p+c)(c+1)+c^2\right)}{(c-1)} \frac{1}{c^{p+2}}. \qquad (4.16)$$

The last of our lemmas here enables us to transfer the field locally to its children.

Lemma 4.4. *Given a complex polynomial centered at z_0, i.e.,*

$$f(z) = \sum_{k=0}^{p} a_k(z-z_0)^k, \qquad (4.17)$$

the expansion of $f(z)$ centered at 0 given by

$$f(z) = \sum_{k=0}^{p} b_k z^k \qquad (4.18)$$

is obtained using Horner's method.

Proofs of the above lemmas, i.e., Lemmas 4.1–4.4, can be found in [26, 27].

Now if we look at the $m\log m$ algorithm, the reason it had that complexity is due to the fact that at each level, the cost was $\mathcal{O}(m)$, since all the particles were directly "accessed" at all the $\log m$ levels. If we are able to avoid this, and reduce the number of particle "accessed" at every level by a factor, then we would get an $\mathcal{O}(m)$ scheme. Lemmas 4.2 and 4.4 enable us to do precisely that.

Lemma 4.2 enables us to obtain a multipole expansion of a cluster using the multipole expansion of its children without "accessing" every particle of the children, while Lemma 4.4 enables us to transfer the field of the cluster due to well-separated clusters to its children. Note that both operations cost us only $4 \times \mathcal{O}(p^2)$ computations. Hence, at level ℓ, since there are $\frac{m/p}{4^\ell}$ clusters, the cost is $\mathcal{O}\left(\frac{m/p}{4^\ell} \times 4p^2\right) = \mathcal{O}\left(p\frac{m}{4^{\ell-1}}\right)$ at level ℓ. Hence, the total cost is $\mathcal{O}(pm)$. The classical FMM is presented in Algorithm 2. There are many further optimizations that could be done to reduce the leading-order coefficient. We refer the reader to [28, 34, 52] for more details.

4.2 Black-Box Fast Multipole Method

In this section, we describe a black-box FMM for nonoscillatory kernels discussed in [21]. The black-box FMM is based on Chebyshev interpolation. The key ingredient behind any interpolation-based FMM is the fact that the interaction between well-separated clusters is of low rank, and this low-rank decomposition can be well represented by interpolating functions. If $w_l(x)$ are the interpolating functions, then we have

---

Algorithm 2 The classical fast multipole method

---

1. Consider a balanced 2D FMM tree with $\kappa = \log_4(m)$ levels.
2. At the leaf level, i.e., at level κ, there are on average p particles per box. Hence, the number of clusters at the leaf level is $\mathcal{O}(m/p)$.
3. At the leaf level, form the multipole expansions for all the clusters using the sources.
4. For levels $\ell \in \{\kappa - 1, \kappa - 2, \ldots, 0\}$, form the multipole expansion for each cluster from its children using Lemma 4.2. This is termed the multipole-to-multipole operation and denoted by M2M.
5. Using Lemma 4.3, compute the local field from the interaction list for all the clusters starting from the root level, i.e., level 0, and proceed till the leaf level, i.e., level κ, is reached. This is termed the multipole-to-local operation and denoted by M2L.
6. Now that we have local expansions for all the clusters at all levels, start at the root and shift the local expansions to its children making use of Lemma 4.4. This is termed the local-to-local operation and denoted by L2L.
7. Once all the above are done, compute the self and neighbor interactions for all the clusters at the leaf level directly.
The total computational complexity of the above algorithm is approximately $40pm$.

$$K(x,y) \approx \sum_l \sum_j K(x_l, y_j) w_l(x) w_j(y). \tag{4.19}$$

The advantage of this type of approach is that there is hardly any precomputing. Also, this works for kernels that are known analytically or defined purely numerically. The low-rank construction depends only on the ability to compute the kernel at a set of nodes and does not rely on any kernel-dependent analytic expansion. One downside of this algorithm is the fact that the low-rank construction, in general, is suboptimal. The black-box FMM relies on Chebyshev interpolation to obtain the low-rank approximation of well-separated clusters. This FMM also uses the singular value decomposition to compress the multipole-to-local operator, which is the bottleneck in most of the FMM implementations. We will explain this in detail as we proceed.

4.2.1 Low-Rank Construction Using Chebyshev Polynomials

A matrix–vector product is nothing but a sum of the form

$$f(x_j) = \sum_{k=1}^{m} K(x_j, y_k) \sigma_k, \tag{4.20}$$

where $j \in \{1, 2, \ldots, m\}$, x_j are the target points, y_k are the source points, and σ_k are the sources and $K(x,y)$ is a kernel. It is precisely sums of these forms in which we are interested, since the covariance matrix Q arises from a kernel function. A direct computation has a computational complexity of $\mathcal{O}(m^2)$, which is prohibitively expensive for large m. However, the computation can be accelerated if the kernel can be well represented by a low-rank approximation, i.e., if

$$K(x,y) \approx \sum_{l=1}^{p} u_l(x)v_l(y), \tag{4.21}$$

where $p \ll m$. Once we have such a low-rank representation, a fast summation technique is evident. We then have

$$f(x_j) \approx \sum_{k=1}^{m} \sum_{l=1}^{p} u_l(x)v_l(y_k)\sigma_k = \sum_{l=1}^{p} u_l(x)\left(\sum_{k=1}^{m} v_l(y_k)\sigma_k\right). \tag{4.22}$$

Hence, the algorithm above scales as $\mathcal{O}(pm)$ as opposed to the $\mathcal{O}(m^2)$ scaling of

---

Algorithm 3 Fast summation when the kernel is well approximated by a low-rank representation

1. Transform the sources using the basis functions $v_l(y)$.

$$W_l = \sum_{k=1}^{m} v_l(y_k)\sigma_k, \text{ where } l \in \{1,2,\dots,p\}.$$

The computational cost of this step is $\mathcal{O}(pm)$.

2. Compute $f(x)$ at target location using the basis function $u_l(x)$.

$$f(x_j) = \sum_{l=1}^{p} u_l(x_j)W_l, \text{ where } j \in \{1,2,\dots,m\}.$$

The computational cost of this step is $\mathcal{O}(pm)$.

---

the direct computation, since $p \ll m$.

Hence, the goal is to construct a low-rank representation, since it allows us to reduce the computational cost of matrix–vector products significantly. The low-rank approximation of a kernel can be constructed by any interpolation scheme, as shown below.

Consider interpolating a function $g(x)$ on an interval $[a,b]$. For pedagogical reasons, let us fix the interval to be $[-1,1]$. An n-point interpolant approximating the function $g(x)$ is given by

$$g_{n-1}(x) = \sum_{k=1}^{n} g(x_k)w_k(x), \tag{4.23}$$

where the x_k are the interpolation nodes and the $w_k(x)$ are the interpolating functions, $k \in \{1,2,\dots,n\}$. The above extends to functions (kernels) $K(x,y)$, which are functions of two variables.

First, by treating $K(x,y)$ as a function of x, we get that

$$K(x,y) \approx \sum_{j=1}^{p} K(x_j,y)w_j(x). \tag{4.24}$$

Now treating $K(x_j, y)$ as a function of y, we get that

$$K(x,y) \approx \sum_{j=1}^{p} \sum_{k=1}^{p} K(x_j, y_k) w_j(x) w_k(y). \tag{4.25}$$

This gives us the desired low-rank representation for the kernel $K(x,y)$, since we have

$$K(x,y) = \sum_{j=1}^{p} u_j(x) v_j(y), \tag{4.26}$$

where $u_j(x) = w_j(x)$ and $v_j(y) = \sum_{k=1}^{p} K(x_j, y_k) w_k(y)$. Note that the above low-rank construction works with any interpolation technique.

One of the most popular interpolation techniques is to interpolate using Chebyshev polynomials, with the Chebyshev polynomials serving as interpolation basis and the Chebyshev nodes serving as interpolation nodes. Let us briefly review some of the properties of Chebyshev polynomials before proceeding further.

The Chebyshev polynomial of the first kind of degree p, denoted by $T_p(x)$, is defined for $x \in [-1, 1]$ and is defined by the recurrence

$$T_p(x) = \begin{cases} 1 & \text{if } p = 0, \\ x & \text{if } p = 1, \\ 2xT_{p-1}(x) - T_{p-2}(x) & \text{if } p \in \mathbb{Z}^+ \backslash \{1\}. \end{cases} \tag{4.27}$$

The polynomial $T_p(x)$ has p distinct roots in the interval $[0, 1]$ located at $\bar{x}_k = \cos\left(\dfrac{2k-1}{2p}\right)$, where $k \in \{1, 2, \ldots, p\}$. These p roots are called Chebyshev nodes.

There are many advantages to Chebyshev interpolation, the greatest one being that it is a stable interpolation scheme, i.e., it doesn't suffer from Runge's phenomenon, and the convergence is uniform.

Hence, the Chebyshev interpolation of a function $g(x)$ is an interpolant of the form

$$g_{p-1}(x) = \sum_{k=0}^{p-1} c_k T_k(x), \tag{4.28}$$

where

$$c_k = \begin{cases} \dfrac{2}{p} \sum_{j=1}^{p} g(\bar{x}_j) T_k(x_j) & \text{if } k > 0, \\ \dfrac{1}{p} \sum_{j=1}^{p} g(\bar{x}_j) & \text{if } k = 0, \end{cases} \tag{4.29}$$

where $\bar{x}_j$ are the Chebyshev nodes, i.e., the p roots of $T_p(x)$. Rearranging the terms, we get that

$$g_{p-1}(x) = \sum_{j=1}^{p} g(\bar{x}_j) S_p(\bar{x}_j, x), \tag{4.30}$$

where

$$S_p(x,y) = \frac{1}{p} + \frac{2}{p} \sum_{k=1}^{p-1} T_k(x) T_k(y). \tag{4.31}$$

If $g(x) \in C^r([-1,1])$, then the error in the approximation is given by

$$|g(x) - g_p(x)| = \mathcal{O}(p^{-r}). \tag{4.32}$$

Further, if the function $g(x)$ can be extended to a function $g(z)$ that is analytic inside a simple closed curve Γ that encloses the points x and all the roots of the Chebyshev polynomial $T_{p+1}(x)$, then the interpolant $g_p(x)$ can be written as

$$g_p(x) = \frac{1}{2\pi i} \oint_{\Gamma} \frac{(T_{p+1}(z) - T_{p+1}(x)) g(z)}{T_{p+1}(z)(z-x)} dz. \tag{4.33}$$

This gives us an error as shown in Eq. (4.6):

$$g(x) - g_p(x) = \frac{1}{2\pi i} \oint_{\Gamma} \frac{T_{p+1}(x) g(z)}{T_{p+1}(z)(z-x)} dz. \tag{4.34}$$

If $g(z)$ is analytic inside an ellipse E_r given by the locus of points $\frac{1}{2}\left(r \exp(i\theta)\right.$ $+ \frac{1}{r} \exp(-i\theta)\left.\right)$, for some $r > 1$ and $\theta \in [0, 2\pi)$, and if $M = \sup_{E_r} |g(z)| < \infty$, then for every $x \in [-1,1]$, we have exponential convergence, since

$$|g(x) - g_p(x)| \le \frac{(r + r^{-1}) M}{(r^{p+1} + r^{-(p+1)})(r + r^{-1} - 2)}. \tag{4.35}$$

Note that $S_p(\bar{x}_j, x)$ is the interpolating function and $\{\bar{x}_j\}_{j=1}^{p}$ are the interpolation nodes. As discussed earlier, this enables us to get the low-rank representation of the kernel $K(x,y)$ using Chebyshev polynomials as

$$K(x,y) \approx \sum_{j=1}^{p} \sum_{k=1}^{p} K(\bar{x}_j, \bar{y}_k) S_p(\bar{x}_j, x) S_p(\bar{y}_k, y). \tag{4.36}$$

This enables us to compute $f(x_j) = \sum_{k=1}^{m} K(x_j,y_j)\sigma_j$, where $j \in \{1,2,\ldots,m\}$, in $\mathcal{O}(pm)$ as opposed to $\mathcal{O}(m^2)$.

Algorithm 4 Fast summation using Chebyshev interpolation to construct the low-rank representation

---

1. Compute the weights at the Chebyshev nodes $\bar{y}_k$ by anterpolation.

$$W_k = \sum_{j=1}^{m} S_p(\bar{y}_k,y_j)\sigma_j, \text{ where } k \in \{1,2,\ldots,p\}.$$

The computational cost of this step is $\mathcal{O}(pm)$.

2. Compute $f(x)$ at the Chebyshev nodes $\bar{x}_l$.

$$f(\bar{x}_l) = \sum_{k=1}^{p} W_k K(\bar{x}_l,\bar{y}_k), \text{ where } l \in \{1,2,\ldots,p\}.$$

The computational cost of this step is $\mathcal{O}(p^2)$.

3. Compute $f(x)$ at the target location x_j by interpolation.

$$f(x_j) = \sum_{l=1}^{p} f(\bar{x}_l)S_p(\bar{x}_l,x_j) \text{ where } j \in \{1,2,\ldots,m\}.$$

The computational cost of this step is $\mathcal{O}(pm)$.

Since $p \ll m$, the computational cost of the algorithm is $\mathcal{O}(pm)$.

The extension of the above to construct low-rank in higher dimension extends with ease. For instance, in d-dimensions, consider the kernel $K(\mathbf{x},\mathbf{y})$, where $\mathbf{x} = (x_1,x_2,\ldots,x_d)$ and $\mathbf{y} = (y_1,y_2,\ldots,y_d)$. A rank p^d low-rank approximation to the kernel $K(\mathbf{x},\mathbf{y})$ is given by

$$K(\mathbf{x},\mathbf{y}) \approx \sum_{j=1}\sum_{k=1} K(\bar{\mathbf{x}}_j,\bar{\mathbf{y}}_k)R_p(\bar{\mathbf{x}}_j,\mathbf{x})R_p(\bar{\mathbf{y}}_k,\mathbf{y}), \qquad (4.37)$$

where

$$R_p(\mathbf{x},\mathbf{y}) = S_p(x_1,y_1)S_p(x_2,y_2)\ldots S_p(x_d,y_d) \qquad (4.38)$$

and $\bar{\mathbf{x}}_j = (\mathbf{x}_{j_1},\mathbf{x}_{j_2},\ldots,\mathbf{x}_{j_d})$ and $\bar{\mathbf{y}}_k = (\mathbf{y}_{k_1},\mathbf{y}_{k_2},\ldots,\mathbf{y}_{k_d})$ are the Chebyshev nodes in d-dimensions.

4.2.2 The Black-Box FMM

In this section, we look at the black-box FMM based on Chebyshev interpolation. If the kernel contains a singularity (or) discontinuity in the domain, then the kernel cannot be well approximated by a low-rank representation. However, in such cases, as in the case of the classical FMM, low-rank representation is applicable when the source and target clusters are well separated, i.e., Eq. (4.36) approximates the far-field behavior of the kernel $K(x,y)$. As with the classical FMM, we recursively subdivide the domain and make use of the low-rank representation wherever applicable to accelerate the computation. Algorithm 5 describes the black-box FMM in 1D. As with all FMM algorithms, in d dimensions, construct a 2^d tree (i.e., a binary tree in 1D, quad tree in 2D, oct tree in 3D, and so on). For pedagogical reasons, we assume that the tree is balanced with κ levels.

Algorithm 5 The black-box fast multipole method

---

1. For all clusters $\mathscr{C}$ at the leaf level, i.e., at level κ, compute the weights at the Chebyshev nodes $\bar{y}_k$ by anterpolation.

$$W_k^{\mathscr{C}} = \sum_{j=1}^{m} S_p\left(\bar{y}_k^{\mathscr{C}}, \mathbf{y}_j\right) \sigma_j, \text{ where } k \in \{1, 2, \ldots, p\}.$$

2. For all clusters $\mathscr{C}$ at level $\ell \in \{\kappa - 1, \kappa - 2, \ldots, 1, 0\}$, compute the weights at the Chebyshev nodes $\bar{y}_k$ by recursion. This is the multipole-to-multipole operation, denoted by **M2M**.

$$W_k^{\mathscr{C}} = \sum_{\mathscr{C}_c- \text{ child of } \mathscr{C}} \sum_{k_c} S_p\left(y_k^{\mathscr{C}}, y_{k_c}^{\mathscr{C}_c}\right) W_{k_c}^{\mathscr{C}_c}, \text{ where } k \in \{1, 2, \ldots, p\}.$$

3. For all clusters $\mathscr{C}$ at level $\ell \in \{0, 1, 2, \ldots, \kappa\}$, compute the far-field contribution at the Chebyshev nodes $\bar{x}_j^{\mathscr{C}}$ for all the clusters in the interaction list of the cluster $\mathscr{C}$. This is the multipole-to-local operation, denoted by **M2L**.

$$g_j^{\mathscr{C}} = \sum_{\mathscr{C}_I \in \mathcal{I}_{\mathscr{C}}} \sum_k K\left(\bar{x}_j^{\mathscr{C}}, \bar{y}_k^{\mathscr{C}_I}\right) W_k^{\mathscr{C}_I}, \text{ where } j \in \{1, 2, \ldots, p\}.$$

4. For all clusters $\mathscr{C}$ at level 0 and level 1, set $f_k^{\mathscr{C}} = g_k^{\mathscr{C}}$.

5. For all clusters $\mathscr{C}$ at level $\ell \in \{2, 3, \ldots, \kappa - 1, \kappa\}$, compute the far-field effect by interpolating the field from the parent cluster at level $\ell - 1$. This is the local-to-local operation, denoted by **L2L**.

$$f_k^{\mathscr{C}} = g_k^{\mathscr{C}} + \sum_{k_p} f_{k_p}^{\mathscr{C}_p} S_p\left(\bar{x}_k^{\mathscr{C}}, \bar{x}_{k_p}^{\mathscr{C}_p}\right), \text{ where } k \in \{1, 2, \ldots, p\}.$$

6. Compute $f(x_j)$ by interpolating the far-field from the Chebyshev nodes to the target locations and computing the nearby and self interactions directly. Here $\mathscr{C}$ is the containing x_j.

$$f(x_j) = \sum_k f_k^{\mathscr{C}} S_p\left(\bar{x}_k^{\mathscr{C}}, x_j\right) + \sum_{\mathscr{C}_N \in \mathcal{N}_{\mathscr{C}}} \sum_{y_k \in \mathscr{C}_N} K(x_j, y_k)\sigma_k + \sum_{y_k \in \mathscr{C}} K(x_j, y_k)\sigma_k, \text{ where}$$
$$j \in \{1, 2, \ldots, m\}.$$

We can further optimize Algorithm 5. In Algorithm 5, the largest contribution is the M2L operation. This is due to the fact that in d dimensions, there can be at most $6^d - 3^d$ clusters in the interaction list of each cluster. (In 2D, there can be at most 27 clusters in the interaction list of each cluster, and in 3D, there can be at most 189 clusters in the interaction list of each cluster.) Hence, to reduce the cost, it is essential to obtain compact multipole and local expansions. One way to further compress the low-rank approximation is to use singular value decomposition. We refer the reader to [21] for more details.

5 Hierarchical Matrix Approach

In the previous section, we discussed the FMM, which relies on the fact that the well-separated clusters are of low rank. The low-rank approximation is obtained using an analytic approach. The hierarchical matrix approach is an algebraic version of the above to capture the fact that well-separated clusters are of low rank.

5.1 Low-Rank Matrices

Before discussing hierarchical matrices, let us first look at low-rank matrices (Fig. 5) and low-rank algebra, since these are the key ingredients for the hierarchical matrix approach. From a theoretical point of view, when we say that a linear operator has rank p, it means that the range of the linear operator is a p-dimensional space. From a matrix algebra point of view, column rank denotes the number of independent columns of a matrix, while row rank denotes the number of independent rows of a matrix. An interesting and nonobvious fact is that the row rank is same as column rank. Hence, we can in fact talk about *the* rank of a matrix. When we say that a matrix $A \in \mathbb{R}^{n \times n}$ has rank p, what it means is that if we take all vectors $x \in \mathbb{R}^{n \times 1}$, then Ax spans a p-dimensional subspace.

From an applied setting, the rank of a matrix denotes the *information content* of the matrix. The lower the rank, the lower the *information content*. For instance, if the rank of a matrix $A \in \mathbb{R}^{n \times n}$ is 1, the matrix can be written as a product of a column vector and a row vector, i.e.,

$$
\begin{pmatrix}
a_{11} & a_{12} & a_{13} & \cdots & a_{1n} \\
a_{21} & a_{22} & a_{23} & \cdots & a_{2n} \\
a_{31} & a_{32} & a_{33} & \cdots & a_{3n} \\
\vdots & \vdots & \vdots & \ddots & \vdots \\
a_{n1} & a_{n2} & a_{n3} & \cdots & a_{nn}
\end{pmatrix}
=
\begin{pmatrix}
1 \\
\alpha_2 \\
\alpha_3 \\
\vdots \\
\alpha_n
\end{pmatrix}
\begin{pmatrix} a_{11} & a_{12} & a_{13} & \cdots & a_{1n} \end{pmatrix},
\tag{5.1}
$$

where $\alpha_k = \dfrac{a_{kj}}{a_{1j}}$. Hence, we need only $2n - 1$ elements to represent an $n \times n$ matrix. So all we need to represent the matrix is $2n - 1$ elements. In general, if we know that a matrix $A \in \mathbb{R}^{m \times n}$ is of rank p, then we can write A as UV^*, where $U \in \mathbb{R}^{m \times p}$, $V \in \mathbb{R}^{n \times p}$, and we need only $2np - p^2$ of its entries. Hence, if we know that a matrix

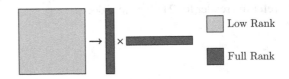

Fig. 5 Outer product representation of a low-rank matrix

is of low rank, then we can compress and store the matrix and perform efficient matrix operations using it. The above ideas can be extended to any linear operator, and these in fact form the basis for various compression techniques.

When the matrix arises out of a kernel function, then analytic techniques such as power series representation and interpolation, as described in Sect. 4, could be used to obtain a low-rank factorization of the matrix. In the next couple of sections, we discuss some exact and approximate algebraic techniques to obtain a low-rank factorization.

5.1.1 Singular Value Decomposition

We first look at the *singular value decomposition* of a matrix. The singular value decomposition (Theorem 5.1) of a matrix is one of the most important decompositions of a matrix and is of much practical importance.

Theorem 5.1. *Consider a matrix $A \in \mathbb{C}^{m \times n}$. Then there exists a factorization of the form*

$$A = U \Sigma V^*, \tag{5.2}$$

where $U \in \mathbb{C}^{m \times m}$, $V \in \mathbb{C}^{n \times n}$ are unitary matrices and $\Sigma \in \mathbb{R}^{m \times n}$ is a diagonal matrix with nonnegative entries on the diagonal, i.e., $\sigma_k = \Sigma_{kk} \geq 0$. Further, $\sigma_1 \geq \sigma_2 \geq \sigma_3 \geq \cdots \geq \sigma_{\min(m,n)} \geq 0$:

$$\Sigma = \begin{pmatrix} \sigma_1 & 0 & 0 & \cdots & 0 \\ 0 & \sigma_2 & 0 & \cdots & 0 \\ 0 & 0 & \sigma_3 & \cdots & 0 \\ \vdots & \vdots & \vdots & \ddots & \vdots \\ 0 & 0 & \vdots & \vdots & \ddots \end{pmatrix}. \tag{5.3}$$

These σ_k's are called the singular values of the matrix A. Here V^* denotes the conjugate transpose of the matrix V. Also, a square matrix $M \in \mathbb{C}^{r \times r}$ is unitary if

$$MM^* = M^*M = I, \tag{5.4}$$

where I is the identity matrix. The decomposition in Eq. (5.2) is called the "singular value decomposition."

The singular value decomposition enables us to compute the *optimal* low-rank approximation of a matrix. Theorem 5.2 reveals how the singular value decomposition gives the "optimal" rank-p approximation to a matrix.

Theorem 5.2. *Consider $A \in \mathbb{C}^{m \times n}$, and let $\|\cdot\|$ be a unitarily invariant matrix norm (for instance, the 2-norm and the Frobenius/Hilbert–Schmidt norm). Let $U \Sigma V^*$ be* *the singular value decomposition of the matrix A. Then* $\forall p \in \{1, 2, \ldots, \min\{m, n\}\}$, *we have*

$$\inf_{M \in \mathbb{R}^{m \times n}} \{\|A - M\| \mid \operatorname{rank}(M) \le p\} = \|A - A_p\|, \tag{5.5}$$

where

$$A_p = U \Sigma_p V^* = \sum_{k=1}^{p} \sigma_k u_k v_k^* \tag{5.6}$$

and $\Sigma_k = \operatorname{diag}(\sigma_1, \sigma_2, \ldots, \sigma_p, 0, \ldots, 0)$. *Also,* $u_i, v_i, i = 1, \ldots, k$ *are the first k columns of the matrices U and V respectively.*

In other words, the "optimal" approximation of a matrix of rank at most p is obtained by retaining the first p singular values and vectors of the matrix. In particular, for A_p as defined before, we have

$$\|A - A_p\|_2 = \sigma_{p+1} \quad \text{and} \quad \|A - A_p\|_{\text{Fro}}^2 = \sum_{j=p+1}^{\min\{m,n\}} \sigma_j^2. \tag{5.7}$$

In terms of a relative error of approximation, it is useful to talk about the ε-rank of a matrix. The ε-rank of a matrix A in the norm $\|\cdot\|$ is defined as

$$p(\varepsilon) := \min \left\{ r \mid \frac{\|A - A_r\|}{\|A\|} \le \varepsilon \right\}, \tag{5.8}$$

where $A_r = U \Sigma_r V^*$.

One of the major drawbacks of using the singular value decomposition in fast algorithms to construct low-rank approximations is that it is computationally expensive. The computational cost to construct a singular value decomposition of an $m \times n$ matrix is $\mathcal{O}(mn \min(m, n))$, which is not desirable in constructing low-rank decompositions.

5.1.2 Pseudo-Skeletal or *CUR* Approximation

The pseudo-skeletal approximation [24] is another technique to construct approximate low-rank factorizations. Though pseudo-skeletal approximations do not yield the optimal low-rank factorizations, the computational cost to construct low-rank factorizations using the pseudo-skeletal approach is independent of the size of the matrix and depends only on the rank. This makes it attractive for use in large systems and in the construction of fast algorithms.

A pseudo-skeletal approximation of a matrix $A \in \mathbb{C}^{m \times n}$ is a decomposition of the form CUR, where $C \in \mathbb{C}^{m \times p}$ is a submatrix of A consisting of certain columns of A,

$R \in \mathbb{C}^{p \times n}$ is another submatrix of A consisting of certain rows of A, and $U \in \mathbb{C}^{p \times p}$ is a square matrix. It is to be noted that there are many different CUR decompositions, and depending on the matrix A, certain CUR decompositions are more optimal than others. Here we look at one such decomposition.

Consider a matrix $A \in \mathbb{C}^{m \times n}$ that has exact rank p. Choose p linearly independent rows and columns. Hence, these r rows span the entire m rows, and the p columns span the entire n columns. Without loss of generality, we can assume that the first p rows and p columns are linearly independent; otherwise, premultiply and postmultiply by permutation matrices to get the linearly independent rows and columns respectively to the top and to the left).

Let C denote the first p columns and R denote the first p rows and U denote the $p \times p$ submatrix at the top left. Clearly, $C = \begin{pmatrix} U \\ C_A \end{pmatrix}$ and $R = (U \ R_A)$ and $A = \begin{pmatrix} U & R_A \\ C_A & M \end{pmatrix}$:

$$CU^{-1}R = \begin{pmatrix} U \\ C_A \end{pmatrix} U^{-1} (U \ R_A) = \begin{pmatrix} U \\ C_A \end{pmatrix} (U^{-1}U \ U^{-1}R_A) \tag{5.9}$$

$$= \begin{pmatrix} U \\ C_A \end{pmatrix} (I \ U^{-1}R_A) = \begin{pmatrix} U & UU^{-1}R_A \\ C_A & C_A U^{-1}R_A \end{pmatrix} \tag{5.10}$$

$$= \begin{pmatrix} U & R_A \\ C_A & C_A U^{-1}R_A \end{pmatrix} = A. \tag{5.11}$$

However, in general, if a matrix A is of rank p, it is hard to choose the p rows and columns exactly. One way to circumvent this issue is to choose, say, kp rows and kp columns. The matrix U now is not invertible. Obtain the pseudo-inverse of U and use singular value decomposition on the pseudo-inverse of U to get the rank-p approximation of the pseudo-inverse of U. This along with the matrices C and R gives us the CUR decomposition. The cost of obtaining the CUR decomposition is $\mathcal{O}(p^3)$.

There are several techniques to choose the "best" set of rows and columns, i.e., the submatrices R and C. These include greedy algorithms, adaptive techniques, random sampling techniques, and interpolatory techniques such as using radial basis functions. Also, in the above algorithm to construct the matrix, U, we ensured that the matrix CUR reproduces exactly certain columns and rows of the matrix A. Another popular technique is to compute the matrix U using a least squares technique, i.e., given C and R, find the matrix U such that $\|A - CUR\|$ is minimum. However, in this case, the computational cost to compute U scales as $\mathcal{O}(m+n)$.

5.1.3 Partially Pivoted Adaptive Cross Approximation

Adaptive cross approximation (ACA) [50] is another technique to construct low-rank approximations of dense matrices. ACA can be thought of as a special case of the pseudo-skeletal algorithm. The idea behind cross approximation is based on the result described in [4], which states that supposing a matrix A is well approximated by a low-rank matrix, by a clever choice of p columns $C \in \mathbb{R}^{m \times p}$ and p rows R of the matrix A, we can approximate A with almost the same approximation quality.

Theorem 5.3. *Assume that the matrix $A^{m \times n}$ has an ε-rank at most p_ε. Then there exists a pseudo-skeletal approximation of the form CUR, where $U \in \mathbb{R}^{p \times p}$ is a matrix that contains appropriate coefficients that are calculated from the submatrix of A in the intersection of rows R and columns C such that*

$$\|A - CUR\|_2 \leq \varepsilon(1 + 2\sqrt{p}(\sqrt{m} + \sqrt{n})). \tag{5.12}$$

Though the algorithm is more efficient than the singular value decomposition, it is still computationally expensive, since it chooses the "best" set of rows and columns and hence costs $\mathcal{O}(pmn)$. Several heuristic strategies have been proposed to reduce the complexity of the ACA algorithm. A more computationally efficient way is the *partially pivoted ACA* algorithm 6. The key ingredient behind partially pivoted ACA is that it needs only the individual entries a_{ij} of the matrix as opposed to the entire row or column. The computational complexity of partially pivoted ACA is $\mathcal{O}(p^2(m+n))$, and it is very easy to implement. A practical version of the algorithm, which includes a termination criterion based on a heuristic approximation to the relative approximation in the Frobenius norm, can be found in [9,51]. The algorithm is described in Algorithm 6. A proof of convergence of this algorithm can be found in [6]. It relies on approximation bounds of Lagrange interpolation and a geometric assumption on the distribution of points, which may not be very practical. For instance, the work [9] lists some contrived counterexamples that show that this algorithm can produce bad pivots. To fix this issue, several other variants have been proposed such as improved ACA (ACA+) and hybrid cross approximation (HCA).

5.2 Low-Rank Algebra

We now look at operations involving low-rank matrices. Throughout this section, we assume that we have the rank-p decomposition of the matrix $A \in \mathbb{R}^{m \times n}$ as UV^T, where $U \in \mathbb{R}^{m \times p}$ and $V \in \mathbb{R}^{n \times p}$. Algorithm 7 represents an efficient matrix–vector product when the matrix is in low-rank form. Algorithm 8 discusses compressing a rank-p matrix into a rank-r matrix, where $r < p$. Algorithms 9 and 10 discuss efficient ways to add and multiply low-rank matrices.

Algorithm 6 Partially pivoted adaptive cross approximation technique

Initial approximation
$$S := 0, \qquad i_1 = 1,$$

for $k = 0, 1, 2, \ldots$ **do**

1. Generation of the row
$$a = A^T e_{i_{k+1}}$$

2. Row of the residuum and the pivot column
$$(R_k)^T e_{i_{k+1}} = a - \sum_{l=1}^{k} (u_k)_{i_{k+1}} v_k \qquad (5.13)$$

$$j_{k+1} = \arg\max_j |(R_k)_{i_{k+1} j}|$$

3. Normalizing constant
$$\gamma_{k+1} = \left((R_k)_{i_{k+1} j_{k+1}} \right)^{-1}$$

4. Generation of the column
$$a = A e_{j_{k+1}}$$

5. Column of the residual and the pivot row
$$R_k e_{j_{k+1}} = a - \sum_{l=1}^{k} (v_k)_{j_{k+1}} u_k \qquad (5.14)$$

$$i_{k+2} = \arg\max_i |(R_k)_{i j_{k+1}}|$$

6. New approximation
$$S_{k+1} = S_k + u_{k+1} v_{k+1}^T$$

end for

Algorithm 7 Matrix–vector product Ax when the matrix is given in low-rank form $A = UV^T$

1. Compute $w = V^T x$. The computational cost of this step is $\mathcal{O}(pn)$
2. Compute $y = Uw$. The computational cost of this step is $\mathcal{O}(pm)$
The overall cost is $\mathcal{O}(p(m+n))$ as opposed to a direct matrix–vector product, which would scale as $\mathcal{O}(mn)$. The reduction in computational cost is significant when $p \ll \min(m,n)$.

Algorithm 8 Compressing a rank-p matrix $A = UV^T$ into a rank-r matrix, where $r < p$

---

1. Compute the QR decomposition of U and V, i.e., $U = Q_U R_U$ and $V = Q_V R_V$. Computational cost is $\mathcal{O}(p^2(m+n))$.
2. Compute the singular value decomposition of $R_U R_V^T$, i.e., $R_U R_V^T = \hat{U}\hat{\Sigma}\hat{V}$. Computational cost is $\mathcal{O}(p^3)$.
3. Set $\tilde{U} = Q_U \hat{U}(:,1:r)$, $\tilde{V} = Q_V \hat{V}(:,1:r)$ and $\tilde{\Sigma} = \hat{\Sigma}(1:r,1:r)$. Computational cost is $\mathcal{O}(rp(m+n))$ in MATLAB notation.
Now $\tilde{A} = \tilde{U}\tilde{\Sigma}\tilde{V}$ is the desired low-rank approximation of UV^T.

Algorithm 9 Adding two low-rank matrices

Given $A = U_A V_A^T$, $B = U_B V_B^T$, where $U_A \in \mathbb{R}^{m \times p}$, $V_A \in \mathbb{R}^{n \times p}$, $U_B \in \mathbb{R}^{m \times r}$ and $V_B \in \mathbb{R}^{n \times r}$.
$A + B = U_A V_A^T + U_B V_B^T = [U_A; U_B][V_A; V_B]^T$ in MATLAB notation.

Algorithm 10 Multiplying two low-rank matrices

Given $A = U_A V_A^T$, $B = U_B V_B^T$, where $U_A \in \mathbb{R}^{m \times p}$, $V_A \in \mathbb{R}^{l \times p}$, $U_B \in \mathbb{R}^{l \times r}$, and $V_B \in \mathbb{R}^{n \times r}$.
Compute $\Sigma_{AB} = V_A^T U_B$. The computational cost of this step is $\mathcal{O}(prl)$.
Now $AB = U_A \Sigma_{AB} V_B^T$ is the desired product in low-rank form.

5.3 Hierarchical Matrices

Hierarchical matrices, introduced by Hackbusch et al. [29–32], are computationally efficient data-sparse approximations of a class of dense matrices. The computational cost to store a typical $m \times m$ dense matrix is $\mathcal{O}(m^2)$. However, the computational cost to store hierarchical matrices is only $\mathcal{O}(pm\log^\alpha m)$, where $\alpha \in \{0,1\}$. The reduction in the storage cost is due to the fact that most of the dense matrices arising out of applications can be subdivided recursively, based on a tree structure, into a hierarchy of rectangular blocks in which most of the submatrices can be efficiently represented as low-rank matrices with rank p. The major advantage of hierarchical matrices is that the computational cost of matrix operations such as matrix–matrix product, matrix decompositions, and solving linear systems is $\mathcal{O}(p^\beta m \log^\gamma m)$, where $\beta, \gamma \in \{1,2\}$. We refer the reader to [1,53,54], where hierarchical matrices have been used in the context of large-scale geostatistical applications.

5.3.1 Hierarchical Semiseparable Matrices

We will focus our attention on a 1D example with a nonsingular kernel that will help illustrate some of the features of hierarchical matrices. Let us, for example, consider the kernel $\kappa_\alpha : [0,1] \times [0,1] \to \mathbb{R}$ defined as

$$\kappa_\alpha(x,y) = \frac{1}{|x-y|+\alpha}, \qquad \alpha > 0. \tag{5.15}$$

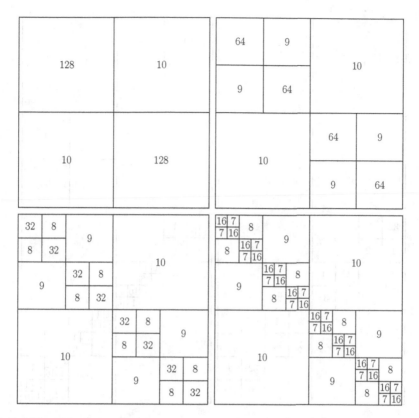

Fig. 6 $k(\varepsilon)$ ranks of the subblocks of the matrix corresponding to $\frac{1}{|x-y|+\alpha}$ in Eq. (5.15), with $\varepsilon = 10^{-6}$ and $m = n = 256$

Notice that this is an appropriate covariance function, even and positive definite (which one can verify by taking the Fourier transform). Let A be an $m \times n$ matrix with entries $A_{ij} = \kappa_\alpha(x_i, y_j)$ for points x_i, y_j discretized in the domain $[0, 1] \times [0, 1]$ with $i = 1, \ldots, m$ and $j = 1, \ldots, n$. Clearly, because this matrix is dense, storing this matrix and computing matrix–vector products are of $\mathcal{O}(mn)$ complexity.

Figure 6 provides a key insight into the structure of the matrix that we can exploit. Again, for the sake of illustration, we consider $m = n = 256$ (even though in actual applications in which hierarchical matrix techniques are useful, m and n are much larger) and compute the ε-ranks of various subblocks of the matrix, for $\varepsilon = 10^{-6}$ and $\alpha = 10^{-6}$. From the first diagram in Fig. 6, we see that the matrix has full rank, which is not surprising. In the second diagram, the block ranks of the 2×2 blocks are provided. We see that the off-diagonal blocks have low ranks, whereas the $(1, 1)$ and $(2, 2)$ blocks still have full rank. We subdivide these blocks further. We see that this kind of hierarchical separation of blocks and approximation of subblocks chosen by some criterion (in this case, simply off-diagonal subblocks) results in reduction in storage requirements and, as we shall see, significant reduction in computational

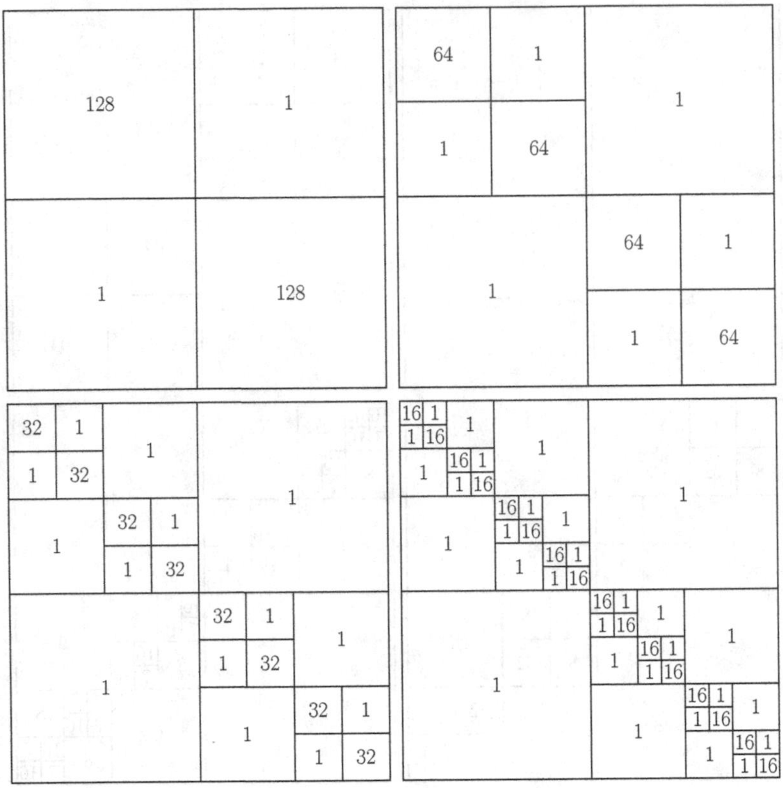

Fig. 7 $k(\varepsilon)$ ranks of the subblocks of the matrix corresponding to $\exp(-|x-y|)$ in Eq. (5.16), with $\varepsilon = 10^{-6}$ and $m = n = 256$

requirements. The amount of memory (in kilobytes) required for storing each level of approximation is as follows: 512, 296, 204, and 122. Clearly, even using just two levels of compression, we can reduce the storage to less than half. This represents a significant reduction in storage for problems of size $\sim 10^6$. As a second example, we consider the exponential covariance kernel, which is part of the Matérn covariance family and is defined as

$$\kappa_l(x,y) = \exp(-|x-y|). \tag{5.16}$$

As before, we plot the ε-ranks of various subblocks of the matrix A with entries $A_{ij} = \kappa(x_i, y_j)$ for $\varepsilon = 10^{-6}$. The points x_i, $i = 1, \ldots, m$, and y_j, $j = 1, \ldots, n$, for $m = n = 256$ are the same as before. However, the situation is much more dramatic. The off-diagonal blocks have rank 1. Figure 7 summarizes the ε-rank of the various subblocks for various levels of approximation. The amount of memory (in kilobytes) required for storing each level of approximation is as follows: 512, 260, 136, and 76. At the finest level, the storage amount is less than 15% of the coarsest level of refinement. We are now in a position to summarize the key features of hierarchical matrices:

- A hierarchical separation of space.
- An acceptable tolerance ε that is specified.
- Low-rank approximation of carefully chosen subblocks.

In the above example, the situation was rather simple because clustering of points can be easily accomplished by grouping them by intervals, and adjacent clusters were of low rank, since the kernel was not discontinuous and had no singularity. The form of recursive low-rank representation of off-diagonal blocks is said to be in the form of hierarchically off-diagonal low-rank matrices. The hierarchical semiseparable matrices (HSS) introduced by Chandrasekaran et al. [11,12,57] are a special class of these matrices. These matrices were introduced as a generalization of semiseparable matrices. A semiseparable matrix [56] of separability rank p is a matrix that can be written as a sum of the lower triangular part of a rank-p matrix and the upper triangular part of another rank-p matrix. Broadly speaking, an HSS matrix is one whose semiseparable structure is hierarchical.

The recursive hierarchical structure of the HSS representation is seen when we consider a 4×4 block partitioning of an HSS matrix $K \in \mathbb{R}^{m \times m}$. The two-level HSS representation is shown in Eq. (5.17):

$$
K = \begin{bmatrix} K_1^{(1)} & U_1^{(1)} B_{1,2}^{(1)} V_2^{(1)^T} \\ U_2^{(1)} B_{2,1}^{(1)} V_1^{(1)^T} & K_2^{(1)} \end{bmatrix}
$$

$$
= \begin{bmatrix} \begin{bmatrix} K_1^{(2)} & U_1^{(2)} B_{1,2}^{(2)} V_2^{(2)^T} \\ U_2^{(2)} B_{2,1}^{(2)} V_1^{(2)^T} & K_2^{(2)} \end{bmatrix} & \begin{bmatrix} U_1^{(2)} S_1^{(2)} \\ U_2^{(2)} S_2^{(2)} \end{bmatrix} B_{1,2}^{(1)} \begin{bmatrix} V_3^{(2)} R_3^{(2)} \\ V_4^{(2)} R_4^{(2)} \end{bmatrix}^T \\ \begin{bmatrix} U_3^{(2)} S_3^{(2)} \\ U_4^{(2)} S_4^{(2)} \end{bmatrix} B_{2,1}^{(1)} \begin{bmatrix} V_1^{(2)} R_1^{(2)} \\ V_2^{(2)} R_2^{(2)} \end{bmatrix}^T & \begin{bmatrix} K_3^{(2)} & U_3^{(2)} B_{3,4}^{(2)} V_4^{(2)^T} \\ U_4^{(2)} B_{4,3}^{(2)} V_3^{(2)^T} & K_4^{(2)} \end{bmatrix} \end{bmatrix}, \quad (5.17)
$$

where $U_i^{(k)}$, $V_i^{(k)}$ can be constructed from $U_{2i-1}^{(k+1)}$, $U_{2i}^{(k+1)}$ and $V_{2i-1}^{(k+1)}$, $V_{2i}^{(k+1)}$ respectively, i.e., we have

$$
U_i^{(k)} = \begin{bmatrix} U_{2i-1}^{(k+1)} S_{2i-1}^{(k+1)} \\ U_{2i}^{(k+1)} S_{2i}^{(k+1)} \end{bmatrix}, \quad (5.18)
$$

$$
V_i^{(k)} = \begin{bmatrix} V_{2i-1}^{(k+1)} R_{2i-1}^{(k+1)} \\ V_{2i}^{(k+1)} R_{2i}^{(k+1)} \end{bmatrix}. \quad (5.19)
$$

The HSS representation of a matrix K with κ levels consists of the following:

1. The diagonal blocks at the level κ, i.e., $K_i^{(\kappa)} \in \mathbb{R}^{m/2^\kappa \times m/2^\kappa}$, where $i \in \{1,2,3,\ldots,2^\kappa\}$. The cost to store all these blocks is $\mathcal{O}(pm)$.
2. $U_i^{(\kappa)}, V_i^{(\kappa)} \in \mathbb{R}^{m/2^\kappa \times p}$ at the lowest level κ, where $i \in \{1,2,3,\ldots,2^\kappa\}$. The cost to store these blocks is $\mathcal{O}(pm)$.

3. $S_i^{(k)}, R_i^{(k)} \in \mathbb{R}^{p \times p}$ at all levels where $k \in \{2, 3, \ldots, \kappa\}$ and $i \in \{1, 2, 3, \ldots, 2^k\}$. The cost to store these blocks at level k is $\mathcal{O}(p^2 2^k)$. Hence, the total cost across all levels is $\mathcal{O}(p^2 2^\kappa) = \mathcal{O}(pm)$.

4. $B_{2i-1,2i}^{(k)}, B_{2i,2i-1}^{(k)} \in \mathbb{R}^{p \times p}$ at all levels where $k \in \{1, 2, \ldots, \kappa\}$ and $i \in \{1, 2, \ldots, 2^{k-1}\}$. The cost to store these blocks at level k is $\mathcal{O}(p^2 2^k)$. Hence, the total cost across all levels is $\mathcal{O}(p^2 2^\kappa) = \mathcal{O}(pm)$.

Hence, the total cost to store an HSS representation of a matrix K is $\mathcal{O}(pm)$.

Taking advantage of the hierarchical structure by recursion and making use of the low-rank algebra discussed in the previous section, all the matrix operations of HSS matrices can be done in $\mathcal{O}(p^\alpha n)$, where p is the rank of the off-diagonal blocks and $\alpha \in \{1, 2\}$.

In a realistic 3-D scenario, the clustering requires a slightly more complicated data structure such as a tree. In the general case, submatrices corresponding to clusters that are well separated lead to low-rank matrices. This will be discussed in the next section.

5.3.2 $\mathcal{H}$-Matrices

Consider the index set $I = \{1, 2, \ldots, m\}$ corresponding to the points $\{x_i\}_{i=1}^m$. A cluster $\tau \subset I$ is a set of indices corresponding to points that are "close" together in some sense. Let X_τ denote the set of points in cluster τ, i.e., $X_\tau = \{x_i \mid i \in \tau\}$. In Sect. 5.1, we discussed some algorithms for low-rank matrices. In practical applications, representing the entire matrix by a low-rank matrix is not possible. In the previous section, on HSS matrices, we saw that certain subblocks of the HSS matrix can be well represented using low-rank matrices. However, if the dense matrix arises from a kernel and the kernel has discontinuities or singularities, only the subblocks corresponding to well-separated clusters can be well represented by low-rank matrices. In order to partition the set of matrix indices $I \times I$ hierarchically into subblocks, we first recursively subdivide the index set I, which leads to a cluster tree. Interested readers can refer to [9] for a formal definition of the cluster tree.

There are many ways to construct a cluster tree. For instance, the FMM tree, shown in Fig. 8 and discussed in Sect. 4, is one such cluster tree. The corresponding hierarchical matrix structure is seen in Fig. 9.

Another simple algorithm to construct the cluster tree is based on geometric bisection, and is described in Algorithm 11. Briefly, this algorithm is initialized with the cluster corresponding to the index set I. The eigenvector v_1 corresponding to the largest eigenvalue of the covariance matrix C is computed, and it corresponds to the direction of the longest expanse of the cluster. Then, we compute the separation plane that goes through the center of mass of the cluster X and is orthogonal to the eigenvector v_1. Based on the side of the separation plane in which the points lie, the cluster is split into two, more or less equal, children. Recursive subdivision gives rise to a binary cluster tree. Theoretically, each node of the tree is split into two children until the cardinality of the node is 1. In practice, this subdivision is stopped if the cardinality of the node is less than or equal to some threshold parameter $n_{\min} \sim p$,

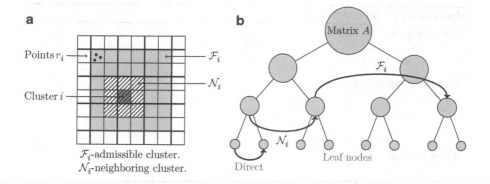

Fig. 8 Fast multipole method tree. (**a**) Hierarchical partition of space based on FMM tree. (**b**) A 2-level FMM cluster tree

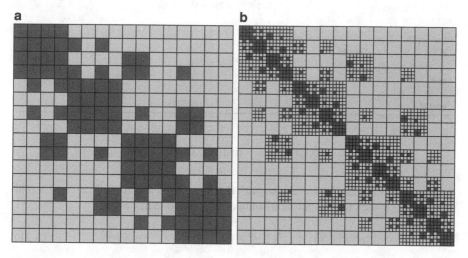

Fig. 9 The hierarchical matrix for a cluster tree based on the 2D fast multipole method. (**a**) Structure of $\mathcal{H}$-matrix at level 2 in the FMM tree. (**b**) Structure of $\mathcal{H}$-matrix at level 3 in the FMM tree

where p is the rank of the interaction between well-separated clusters. For uniformly distributed clusters, the depth of the cluster tree, i.e., the number of levels in the tree, is $\log_2(|I|)$.

Next, well-separated cluster pairs, called *admissible* cluster pairs, are identified. A cluster pair (τ, σ) is considered admissible if

$$\min\{\operatorname{diam}(X_\tau), \operatorname{diam}(X_\sigma)\} \leq \eta \operatorname{dist}(X_\tau, X_\sigma), \tag{5.20}$$

where the diameter of a cluster τ is the maximal distance between any pair of points in τ, and the distance between two clusters τ and σ is the minimum distance between any pair of points in $\tau \times \sigma$

Algorithm 11 Split cluster tree - τ

Center of mass of the cluster

$$X = \frac{1}{m} \sum_{k=1}^{m} x_k \in \mathbb{R}^d$$

Covariance matrix of the cluster

$$C = \sum_{k=1}^{m} (x_k - X)(x_k - X)^T \in \mathbb{R}^{d \times d}$$

Eigenvalues and eigenvectors

$$Cv_i = \lambda_i v_i, \qquad i = 1,\dots,d, \qquad \lambda_1 \geq \cdots \geq \lambda_d \geq 0$$

Initialize

$$\tau_1 := \emptyset, \qquad \tau_2 := \emptyset$$

for $k = 1,\dots,m$
if $(x_k - X, v_1) \geq 0$ **then**
    $\tau_1 := \tau_1 \bigcup x_k$
else
    $\tau_2 := \tau_2 \bigcup x_k$
end if

$$\text{diam}(X_\tau) = \max_{x,y \in X_\tau} \|x - y\| \qquad \text{dist}(X_\tau, X_\sigma) = \min_{x \in X_\tau, y \in X_\sigma} \|x - y\|.$$

A kernel is called *asymptotically smooth* if there exist constants c_1^{as}, c_2^{as} and a real number $g \geq 0$ such that for all multi-indices $\alpha \in \mathbf{N}_0^d$, one has

$$\left| \partial_y^\alpha \kappa(\mathbf{x}, \mathbf{y}) \right| \leq c_1^{as} p! (c_2^{as})^p (|\mathbf{x} - \mathbf{y}|)^{-g-p}, \qquad p = |\alpha|. \tag{5.21}$$

Equation (5.20) ensures that the kernel $\kappa(\cdot, \cdot)$ is asymptotically smooth over the domain $D_\tau \times D_\sigma$, where D_σ is the convex hull of X_σ. Asymptotically smooth kernel functions can be shown to admit degenerate approximations on the kernel functions, on pairs of domains satisfying the admissibility condition (5.20). The implication of the kernel being asymptotically smooth on admissible clusters is that the submatrix corresponding to the blocks $\tau \times \sigma$ have exponentially decaying singular values and are well approximated by low-rank matrices. To see this, we compute the Taylor series of the covariance kernel around a point $y_0 \in D_\sigma$, where D_σ is the convex hull of X_σ:

$$\kappa(x,y) = \sum_{l=0}^{k-1} \frac{(y - y_0)^l}{l!} \partial_y^l \kappa(x, y_0) + R_k(x, y), \tag{5.22}$$

where $R_k(x, y)$ is the remainder in the Taylor series expansion.

The low rank of admissible clusters can be explained by appealing to the Taylor series representation of the kernel κ_α. Consider two sets of indices t, s, and let $X_t = \{x_i \mid i \in t\}$ and $X_s = \{y_j \mid j \in s\}$ satisfy the following condition:

$$\min\{\mathrm{diam}(X_t), \mathrm{diam}(X_s)\} \le \eta \ \mathrm{dist}(X_t, X_s), \qquad 0 < \eta < 1, \qquad (5.23)$$

where we define

$$\mathrm{diam}(X) = \max_{x,y \in X} |x - y|, \qquad \mathrm{dist}(X, Y) = \min_{x \in X, y \in Y} |x - y|. \qquad (5.24)$$

Now using the Taylor series expansion for $\kappa_\alpha(x, y)$ around $\tilde{y} := \max_{X_s} y$ gives us

$$\kappa_\alpha(x, y) = \sum_{i=0}^{\infty} \frac{(y - \tilde{y})^i}{i!} \partial_y^i \kappa_\alpha(x, \tilde{y}) \qquad (5.25)$$

$$= \sum_{i=0}^{k-1} \frac{(y - \tilde{y})^i}{i!} \partial_y^i \kappa_\alpha(x, \tilde{y}) + R_k(x, y), \qquad (5.26)$$

where $R_k(x, y)$ is the remainder term in the Taylor series expansion, which can be bounded as follows:

$$R_k(x, y)| \le \frac{1}{k!} d^k \left(\frac{|y - y_0|}{|x - \tilde{y}|} \right)^k c_1 k! c_2^k |x - \tilde{y}|^{-g} \qquad (5.27)$$

$$\le c_1^{as} \mathrm{dist}^g(X_\tau, X_\sigma)(c_2^{as} d\eta)^k := \varepsilon_k, \qquad (5.28)$$

provided that $\mathrm{diam}(X_s) \le \eta \ \mathrm{dist}(X_t, X_s)$ and α is small. Repeating the same argument with the roles of x and y interchanged, a similar result can be obtained with $\mathrm{diam}(X_t) \le \eta \ \mathrm{dist}(X_t, X_s)$.

The cluster tree can then be used to define a block cluster tree, by forming pairs of clusters recursively. Given a cluster tree and an admissibility condition (5.20), the block cluster tree can be constructed using Algorithm 12. By calling the algorithm with $\tau = \sigma = I$, we create a block cluster tree with root $I \times I$. The leaves of the block cluster tree form a partition of $I \times I$.

Algorithm 12 BuildBlockTree($\tau \times \sigma$)

---

if $\tau \times \sigma$ is not admissible **and** $|\tau| > n_{\min}$ **and** $|\sigma| > n_{\min}$ **then**
    $S(\tau \times \sigma) := \{\tau' \times \sigma' : \tau' \in S(\tau), \sigma' \in S(\sigma)\}$
    for all $\tau' \times \sigma' \in S(\tau \times \sigma)$ **do**
        BuildBlockTree($\tau' \times \sigma'$)
    end for
else
    $S(\tau \times \sigma) := \varnothing$
end if

For the leaves of the block-cluster tree, i.e., those nodes that do not have any children, if the clusters are not admissible, then we store the matrix corresponding to this subblock in a componentwise fashion. Otherwise, if they are admissible, then the admissibility condition (5.20) guarantees that the subblocks will be of low rank, which can be computed in one of the methods described in Sect. 5.1, and the corresponding low-rank matrices will be stored in the outer-product form described in Sect. 5.1.

5.4   Complexity Estimates

Now we define the set of hierarchical matrices with blockwise rank k.

Definition 5.1. Let $\mathcal{T}_I$ be a cluster tree and $\mathcal{T}_{I \times I}$ the block cluster tree corresponding to the index set I. We define the set of hierarchical matrices as

$$\mathcal{H}(\mathcal{T}_{I \times I}, k) := \{ M \in \mathbb{R}^{|I| \times |I|} \,|\, \mathrm{rank}(M_{\tau \times \sigma}) \le k \quad \forall \quad \tau, \sigma \in \mathcal{T}_I \}. \tag{5.29}$$

Note that $\mathcal{H}(\mathcal{T}_{I \times I}, k)$ is not a linear space. For the subblock $A_{\tau \times \sigma} \in \mathbb{R}^{\tau \times \sigma}$, which is the restriction of the matrix A to the subblock $\tau \times \sigma$, and supposing that this cluster pair is admissible, i.e., that it satisfies condition (5.20), it is possible to generate a low-rank approximation for this subblock when the kernel is asymptotically smooth (for the definition, see [5]). The low-rank representation for this subblock is computed using the partially pivoted ACA algorithm [5, 51].

To compute the matrix–vector product involving the $\mathcal{H}$−matrix A with a vector x, we use Algorithm 5.4.

Algorithm 13 Matrix–Vector Product MVM$(A, \tau \times \sigma, x, y)$

---

if $S(\tau \times \sigma) \ne 0$ **then**
    for all $\tau' \times \sigma' \in S(\tau \times \sigma)$ **do**
        $MVM(A, \tau' \times \sigma', x, y)$
    end for
else
    $y|_\tau := y|_\tau + A_{\tau \times \sigma} x_\sigma$, for both full and low-rank subblocks.
end if

The complexity estimates for the operations in the construction and matrix–vector products have been established in several references [5, 25]. They make use of an idea known as the sparsity constant, which is independent of $|I|$ and which we define below as follows.

Definition 5.2 (Sparsity Constant). Let $T_{I \times I}$ be a block-cluster tree corresponding to the cluster tree T_I. The sparsity constant C_{sp} of $T_{I \times I}$ is defined as

$$C_{sp} = \max\{\max_{r \in T_I} \#\{s \in T_I | r \times s \in T_{I \times I}\}, \max_{s \in T_I} \#\{r \in T_I | r \times s \in T_{I \times I}\}\}. \quad (5.30)$$

With this definition, we can then summarize the complexity estimates of storage and matrix–vector products.

Theorem 5.4. *Let $T_{I \times I}$ be a block-cluster tree corresponding to the cluster tree T_I with sparsity constant C_{sp} and depth p. The storage requirement $N_{st}(T_{I \times I}, k)$ for an $\mathcal{H}$-matrix $\mathcal{H}(T_{I \times I}, k)$ is given by*

$$N_{st}(T_{I \times I}, k) \leq 2C_{sp}(p+1)\max\{k, n_{\min}\}|I|, \quad (5.31)$$

and the computational cost of the matrix–vector product $N_{\mathcal{H}x}(T_{I \times I}, k)$ is bounded by

$$N_{\mathcal{H}x}(T_{I \times I}, k) \leq N_{st}(T_{I \times I}, k). \quad (5.32)$$

For more details the reader is referred to [5, 25].

Another data-sparse representation is the class of $\mathcal{H}^2$-matrices, which is a subset of the $\mathcal{H}$-matrix and is very similar to the matrix structure arising out of the FMM cluster-tree structure. The computational cost for storing these $\mathcal{H}^2$ matrices is $\mathcal{O}(pm)$, and the cost for matrix operations is $\mathcal{O}(p^\alpha m)$, where $\alpha \in \{1, 2\}$.

6 Conclusions

This article reviews a few fast algorithms applicable for large-scale stochastic inverse modeling. Since the covariance matrices arise from a nonoscillatory kernel, we look at fast algorithms applicable only when the kernel is nonoscillatory. The algorithms presented in this review are the fast Fourier transform, classical FMM, black-box FMM, and the hierarchical matrix approach. We refer readers to [7, 10, 33, 44, 45] for details on other fast algorithms that can be applied to accelerate computations involving large covariance matrices.

Acknowledgements The authors were supported by "NSF Award 0934596, Subsurface Imaging and Uncertainty Quantification," "Army High Performance Computing Research Center" (AH-PCRC, sponsored by the U.S. Army Research Laboratory under contract No. W911NF-07-2-0027) and "The Global Climate and Energy Project" (GCEP) at Stanford.

References

1. Ambikasaran, S., Li, J., Darve, E., Kitanidis, P.K.: Large-scale stochastic linear inversion using hierarchical matrices (2012). In review
2. Beatson, R., Greengard, L.: A short course on fast multipole methods. Wavelets, multilevel methods and elliptic PDEs pp. 1–37 (1997)

3. Beatson, R., Newsam, G.: Fast evaluation of radial basis functions: I.   Computers & Mathematics with Applications **24**(12), 7–19 (1992)
4. Bebendorf, M.: Approximation of boundary element matrices. Numerische Mathematik **86**(4), 565–589 (2000)
5. Bebendorf, M.: Hierarchical Matrices: A Means to Efficiently Solve Elliptic Boundary Value Problems, *Lecture Notes in Computational Science and Engineering (LNCSE)*, vol. 63. Springer-Verlag (2008). ISBN 978-3-540-77146-3
6. Bebendorf, M., Rjasanow, S.: Adaptive low-rank approximation of collocation matrices. Computing **70**(1), 1–24 (2003)
7. Beylkin, G., Coifman, R., Rokhlin, V.: Fast wavelet transforms and numerical algorithms I. Communications on Pure and Applied Mathematics **44**(2), 141–183 (1991)
8. Bjørnstad, O., Falck, W.: Nonparametric spatial covariance functions: estimation and testing. Environmental and Ecological Statistics **8**(1), 53–70 (2001)
9. Börm, S., Grasedyck, L., Hackbusch, W.: Hierarchical matrices. Lecture notes **21** (2003)
10. Burrus, C., Gopinath, R., Guo, H.: Introduction to wavelets and wavelet transforms: A primer. Recherche **67**, 02 (1998)
11. Chandrasekaran, S., Dewilde, P., Gu, M., Pals, T., Sun, X., van der Veen, A., White, D.: Some fast algorithms for sequentially semiseparable representations. SIAM Journal on Matrix Analysis and Applications **27**(2), 341 (2006)
12. Chandrasekaran, S., Gu, M., Pals, T.: A fast ULV decomposition solver for hierarchically semiseparable representations. SIAM Journal on Matrix Analysis and Applications **28**(3), 603 (2006)
13. Cheng, H., Greengard, L., Rokhlin, V.: A fast adaptive multipole algorithm in three dimensions. Journal of Computational Physics **155**(2), 468–498 (1999)
14. Coifman, R., Rokhlin, V., Wandzura, S.: The fast multipole method for the wave equation: A pedestrian prescription. Antennas and Propagation Magazine, IEEE **35**(3), 7–12 (1993)
15. Cooley, J., Tukey, J.: An algorithm for the machine calculation of complex fourier series. Math. Comput **19**(90), 297–301 (1965)
16. Cornford, D., Csató, L., Opper, M.: Sequential, Bayesian geostatistics: a principled method for large data sets. Geographical Analysis **37**(2), 183–199 (2005)
17. Darve, E.: The fast multipole method: numerical implementation. Journal of Computational Physics **160**(1), 195–240 (2000)
18. Darve, E.: The fast multipole method - I: Error analysis and asymptotic complexity. SIAM Journal on Numerical Analysis pp. 98–128 (2001)
19. Darve, E., Havé, P.: Efficient fast multipole method for low-frequency scattering. Journal of Computational Physics **197**(1), 341–363 (2004)
20. Darve, E., Havé, P.: A fast multipole method for Maxwell equations stable at all frequencies. Philosophical Transactions of the Royal Society of London. Series A: Mathematical, Physical and Engineering Sciences **362**(1816), 603–628 (2004)
21. Fong, W., Darve, E.: The black-box fast multipole method. Journal of Computational Physics **228**(23), 8712–8725 (2009)
22. Fritz, J., Neuweiler, I., Nowak, W.: Application of FFT-based algorithms for large-scale universal kriging problems. Mathematical Geosciences **41**(5), 509–533 (2009)
23. Golub, G., Van Loan, C.: Matrix computations, vol. 3. Johns Hopkins Univ Press (1996)
24. Goreinov, S., Tyrtyshnikov, E., Zamarashkin, N.: A theory of pseudoskeleton approximations. Linear Algebra and Its Applications **261**(1–3), 1–21 (1997)
25. Grasedyck, L., Hackbusch, W.: Construction and arithmetics of $\mathcal{H}$-matrices. Computing **70**, 2003 (2003)
26. Greengard, L., Rokhlin, V.: A fast algorithm for particle simulations. Journal of Computational Physics **73**(2), 325–348 (1987)
27. Greengard, L., Rokhlin, V.: A new version of the fast multipole method for the Laplace equation in three dimensions. Acta Numerica **6**(1), 229–269 (1997)
28. Greengard, L., Rokhlin, V., SCIENCE., Y.U.N.H.C.D.O.C.: On the efficient implementation of the fast multipole algorithm. Defense Technical Information Center (1988)

29. Hackbusch, W.: A sparse matrix arithmetic based on $\mathcal{H}$-matrices. Part I: Introduction to $\mathcal{H}$-matrices. Computing **62**(2), 89–108 (1999)
30. Hackbusch, W., Börm, S.: Data-sparse approximation by adaptive $\mathcal{H}^2$-matrices. Computing **69**(1), 1–35 (2002)
31. Hackbusch, W., Grasedyck, L., Börm, S.: An introduction to hierarchical matrices. Max-Planck-Inst. für Mathematik in den Naturwiss. (2001)
32. Hackbusch, W., Khoromskij, B.: A sparse $\mathcal{H}$-matrix arithmetic. Computing **64**(1), 21–47 (2000)
33. Hackbusch, W., Nowak, Z.: On the fast matrix multiplication in the boundary element method by panel clustering. Numerische Mathematik **54**(4), 463–491 (1989)
34. Hrycak, T., Rokhlin, V.: An improved fast multipole algorithm for potential fields. Tech. rep., DTIC Document (1995)
35. Kitanidis, P.K.: Statistical estimation of polynomial generalized covariance functions and hydrologic applications. Water Resources Research **19**(4), 909–921 (1983)
36. Kitanidis, P.K.: Generalized covariance functions in estimation. Mathematical Geology **25**(5), 525–540 (1993)
37. Kitanidis, P.K.: Quasi-linear geostatistical theory for inversing. Water Resources Research **31**(10), 2411–2419 (1995)
38. Kitanidis, P.K.: On the geostatistical approach to the inverse problem. Advances in Water Resources **19**(6), 333–342 (1996)
39. Kitanidis, P.K.: Introduction to geostatistics: applications to hydrogeology. Cambridge Univ Pr (1997)
40. Kitanidis, P.K.: Generalized covariance functions associated with the Laplace equation and their use in interpolation and inverse problems. Water Resources Research **35**(5), 1361–1367 (1999)
41. Kitanidis, P.K.: On stochastic inverse modeling. Geophysical Monograph-American Geophysical Union **171**, 19 (2007)
42. Kitanidis, P.K., Vomvoris, E.G.: A geostatistical approach to the inverse problem in groundwater modeling (steady state) and one-dimensional simulations. Water Resources Research **19**(3), 677–690 (1983)
43. Liu, X., Illman, W., Craig, A., Zhu, J., Yeh, T.: Laboratory sandbox validation of transient hydraulic tomography. Water Resources Research **43**(5), W05,404 (2007)
44. Mallat, S.: A theory for multiresolution signal decomposition: The wavelet representation. Pattern Analysis and Machine Intelligence, IEEE Transactions on **11**(7), 674–693 (1989)
45. Mallat, S.: A wavelet tour of signal processing. Academic Press (1999)
46. Nishimura, N.: Fast multipole accelerated boundary integral equation methods. Applied Mechanics Reviews **55**, 299 (2002)
47. Nowak, W., Cirpka, O.A.: Geostatistical inference of hydraulic conductivity and dispersivities from hydraulic heads and tracer data. Water Resources Research **42**(8), 8416 (2006)
48. Nowak, W., Tenkleve, S., Cirpka, O.: Efficient computation of linearized cross-covariance and auto-covariance matrices of interdependent quantities. Mathematical geology **35**(1), 53–66 (2003)
49. Pollock, D., Cirpka, O.: Fully coupled hydrogeophysical inversion of synthetic salt tracer experiments. Water Resources Research **46**(7), W07,501 (2010)
50. Rjasanow, S.: Adaptive cross approximation of dense matrices. IABEM 2002, International Association for Boundary Element Methods (2002)
51. Rjasanow, S., Steinbach, O.: The fast solution of boundary integral equations. Mathematical and Analytical Techniques with Applications to Engineering. Springer, New York (2007)
52. Rokhlin, V.: Rapid solution of integral equations of classical potential theory. Journal of Computational Physics **60**(2), 187–207 (1985)
53. Saibaba, A., Ambikasaran, S., Li, J., Darve, E., Kitanidis, P.: Application of hierarchical matrices to linear inverse problems in geostatistics. Oil & Gas Science and Technology–Revue d'IFP Energies nouvelles **67**(5), 857–875 (2012)

54. Saibaba, A., Kitanidis, P.: Efficient methods for large-scale linear inversion using a geostatistical approach. Water Resources Research **48**(5), W05,522 (2012)
55. Starks, T., Fang, J.: On the estimation of the generalized covariance function. Mathematical Geology **14**(1), 57–64 (1982)
56. Vandebril, R., Barel, M., Golub, G., Mastronardi, N.: A bibliography on semiseparable matrices. Calcolo **42**(3), 249–270 (2005)
57. Xia, J., Chandrasekaran, S., Gu, M., Li, X.: Fast algorithms for hierarchically semiseparable matrices. Numerical Linear Algebra with Applications **17**(6), 953–976 (2010)
58. Ying, L., Biros, G., Zorin, D.: A kernel-independent adaptive fast multipole algorithm in two and three dimensions. Journal of Computational Physics **196**(2), 591–626 (2004)

Modeling Spatial and Structural Uncertainty in the Subsurface

Margot Gerritsen and Jef Caers

Abstract Whether subsurface modeling is performed on a laboratory or reservoir scale, uncertainty is inherently present due to lack of data and lack of understanding of the underlying phenomena and processes taking place. We highlight several issues, techniques, and practical modeling tools available for modeling spatial and structural uncertainty of complex subsurface models. We pay particular attention to the method of training images, since this is a recent and novel approach that we think can make significant advances in uncertainty modeling in the subsurface and other application areas. The subsurface is a difficult area of modeling uncertainty: porosity and permeability of subsurface rock can vary orders of magnitude on a variety of scales, fractures or faults may exist, and the available data can be sparse. Modeling of uncertainty in the subsurface is of much interest to society, because of, for example, the exploration and extraction of natural resources including groundwater, fossil fuels, and geothermal energy; the storage of nuclear material; and sequestration of carbon dioxide. Although we focus on the subsurface, many of the techniques discussed are applicable to other areas of uncertainty modeling in engineering and the sciences.

Keywords Uncertainty • Training Images • Subsurface • Heterogeneity

M. Gerritsen (✉)
Department of Energy Resources Engineering, Institute for Computational & Mathematical Engineering, Stanford University, Stanford, CA 94305, USA
e-mail: margot.gerritsen@stanford.edu

J. Caers
Department of Energy Resources Engineering, Stanford University, Stanford, CA 94305, USA

C. Dawson and M. Gerritsen (eds.), *Computational Challenges in the Geosciences*, The IMA Volumes in Mathematics and its Applications 156, DOI 10.1007/978-1-4614-7434-0_6,
© Springer Science+Business Media New York 2013

1 Modeling Uncertainty: Introduction

In this paper, we focus on modeling uncertainty in the subsurface. Many of the ideas discussed, however, carry through to other disciplines, since often the challenges facing the engineer or scientist are similar across application areas. In the subsurface, as in other areas, the data available to constrain the models of uncertainty are diverse and multiscale. Local data may be available from wells (drillers' logs, well log), while at a larger spatial scale, remote sensing measurements may be available (seismic). Tying all these data into a single model of uncertainty without making too many assumptions about the relationships between various data sources is not straightforward. Another commonality between uncertainty modeling in the subsurface and other areas is that it is often tailored to the application. For example, if one is looking to quantify the global reserves of an oil reservoir, then the focus should be on the structural model and global parameters such as net-to-gross, while if the question is about drilling the next well, than the analysis should focus on local reservoir heterogeneity and connectivity of flow units. Also, typically, uncertainty is not a goal on its own but rather a critical component in decision making, which impacts the approaches used.

An important premise in uncertainty quantification is that uncertainty can never be objectively measured. Any assessment of uncertainty will need to be based on a model, chosen by the engineer or scientist based on knowledge and intuition of the processes under consideration. It is important to remind oneself that a model of uncertainty is indeed just that, a model, and therefore in itself uncertain and subject to change as our understanding of processes improves and observational capabilities increase.

Uncertainty can be introduced through various sources:

- Measurement errors, and errors introduced through processing of raw measurements
- Data interpretation, which is often biased and may be hindered by a lack of understanding
- Geological uncertainty, that is, uncertainty about the type of geological setting to be used, which is interpreted from data or based on physical models that themselves are uncertain
- Spatial uncertainty resulting from data sparsity
- Unavailability of data, because they cannot be measured or are not measured because of prohibitive costs
- Response uncertainty, that is, uncertainty about the level and character of the impact that uncertainty in the available data may have on modeling reservoir processes, such as flow, transport, and wave propagation, and management decisions
- Epistemological uncertainty, which is uncertainty introduced because we do not always know what we do not know.

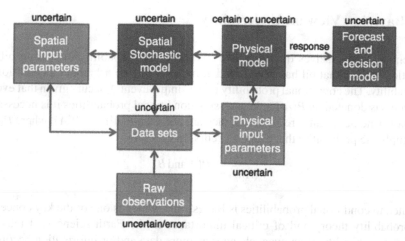

Fig. 1 Overview of uncertainty modeling components in the subsurface

Naturally, such diverse sources for uncertainty complicate uncertainty model development. An additional complication is that the available data exist at various scales, both spatial and temporal. For example, well-core data give us very local information about the reservoir rock properties, whereas seismic data provide information at the full reservoir scale. Another challenge is the very large number of variables needed to describe subsurface processes well. Typically, we will work with gridded models to represent all aspects of the system. Even if each grid cell in a model contains only a couple of variables, we easily have millions of variables for even relatively small models. Unfortunately, probability theory and statistical techniques commonly taught and practiced were not developed with such large and complex systems in mind and are consequently often impractical. The same is true for traditional statistical methods for sensitivity analyses that are often necessary to determine which factors impact our decisions most.

When creating models of uncertainty, the various sources of uncertainty mentioned above are tied together with deterministic (or not) laws of physical and dynamic processes tailored to the scientific, engineering, or management purpose for which the models will be used. Figure 1 provides a general overview of the various components of modeling uncertainty in the subsurface.

In this paper, we will focus on modeling of spatial and structural uncertainties. To lay the foundation, we first discuss modeling of spatial continuity after giving a brief introduction to Bayesian modeling, which provides the framework of the uncertainty modeling in this paper. Other frameworks, e.g., fuzzy logic, are available but are not as mature as bayesian modeling at present.

This review paper is based on the recent book by Caers [9]. We also recommend [5, 6, 8, 10, 11, 14, 16] for information about spatial statistics and geostatistics, and [7, 17, 22] for additional views on modeling uncertainty in the geosciences.

2 Bayesian View on Uncertainty

When we ask ourselves questions like "What is the probability of finding oil at location y given that oil has been found at location x?" we ask about a conditional probability. The conditional probability that a simple event A occurs given that event B occurs is donated by $P(A|B)$. In assessing conditional probabilities it is necessary to know whether events are related. If they are not, then $P(A|B) = P(A)$, where $P(A)$ is simply the probability that event A occurs. In general,

$$P(A|B) = \frac{P(A \text{ and } B)}{P(B)}.$$

Related to conditional probabilities is Bayes' rule, which is one of the key concepts in probability theory and of critical importance in the Earth sciences. It tells us how the probability of an event changes as more data and/or information becomes available. To derive it, we first observe that $P(A \text{ and } B) = P(A|B)P(B)$, but it is also given by $P(A \text{ and } B) = P(B|A)P(A)$. We add these two expressions and divide by 2 to get

$$1 = \frac{P(A|B)P(B)}{P(B|A)P(A)},$$

or

$$P(A|B) = \frac{P(B|A)P(A)}{P(B)}.$$

This equation, which is Bayes' rule, shows how if one conditional probability is known, the other conditional probability can be calculated [12]. Here, $P(A|B)$ is often termed a *posterior* probability (after learning from the data), while $P(A)$ is termed a *prior* probability (before having any data).

Although we considered A and B to be simple events, we can make these events as complex as we want. For example, we can consider A to be the complete unknown reservoir that we would like to model, and B all the available data, possibly given by n distinct data sets B_i, $i = 1, \ldots, n$. Both A and B can be seen as vectors containing the many variables and data to be considered. Here, A may contain variables for each grid cell of a reservoir flow model, which may add up to many millions of elements. Irrespective of the size of A or B, Bayes' rule applies:

$$P(A|B) = P(A|B_1, \ldots, B_n) = \frac{P(B_1, \ldots, B_n|A)P(A)}{P(B_1, \ldots, B_n)} \approx P(B_1, \ldots, B_n|A)P(A).$$

The probability distribution $P(A|B_1, B_2, \ldots, B_n)$ describes the frequency of outcomes of A (if A contains only discrete variables) or the probability density (if A contains continuous variables). If we draw a set of L samples $a_1, a_2, a_3, \ldots, a_L$ from this distribution by Monte Carlo simulation, the set is also a model of uncertainty that approximates P as L goes to infinity. In most cases, we need only draw a few hundred or a few thousand samples to obtain a good approximation. Bayes' rule shows that a model of uncertainty $P(A|B_1, B_2, \ldots, B_n)$, which is also referred to as the *posterior probability*, is fully defined by a *prior model of uncertainty* $P(A)$ and the relationship between data B and model A described by $P(B|A)$, which is termed the likelihood probability, or simply the *likelihood*.

The likelihood can often be determined from physical models and is typically not the modeling bottleneck (although it is a computational bottleneck). Determining prior probabilities is, however, generally difficult and subjective. In a true Bayesian sense, the prior probability (or prior model of uncertainty) is determined without taking into account any data. Of course, this is rarely the case in reality, since modelers typically look at data when performing a study and use it to narrow down model definitions. For example, when modeling a reservoir, it may be clear that the reservoir is a carbonate system, and a priori exclusion of, say, all clastic systems is acceptable. In other words, researchers generally use data twice: once to make hypotheses on the prior, and once to calibrate and condition the model. Prior probabilities can be determined from historical observations if they are deemed good analogues for the study at hand, or by interviewing experts. This procedure is then termed prior uncertainty elicitation from experts, where elements of psychology are brought to the table to make the determination of probabilities as realistic as possible. Whatever method is used, Bayes' rule suggests that at least a mental exercise should be conducted to collect all possibilities imaginable prior to including data into the model. Putting too much focus immediately on data is extremely tempting, since raw observations are the only "hard facts" available, but it is also risky, since eliminating too many possibilities a priori from data that are by nature incomplete and/or erroneous may lead to unrealistically small prior uncertainties.

In a way, any modeling of uncertainty is, particularly in cases with considerable uncertainty, only as good as the modeling of the prior. The prior becomes more important as the complexity of a model grows. If relatively simple physical modeling and few parameters or variables are used in a study, it may be possible to determine unique model parameters deterministically from the available data (for example by inverse modeling). This will lead to a perfect match and a false sense of security. However, often, when more data become available, a mismatch may be observed between this model and the new data. To address this problem, complexity is often added to the model in order to match the new data in addition to the existing data. This is usually done in the belief that the mismatch is caused by an incomplete or incorrect model. However, increasing the complexity of the physical model typically requires more model parameters that cannot be deterministically determined from data. Perhaps counterintuitively, this means that more data often increases rather than decreases the uncertainty in the model.

3 Modeling Spatial Continuity

Modeling spatial continuity is critical to the question of addressing uncertainty. If the spatial properties or values studied are modeled, rather than assumed random, a different assessment of uncertainty will be obtained. Spatial continuity models are particularly important in the Earth sciences because Earth science phenomena are inherently nonrandom and spatial relationships exist. The term "spatial relationship" is interpreted broadly and may incorporate, for example, relationships among the available spatial data or relationships between the unknown properties and the measured data. The data may also be of any type, possibly different from that of the property or value being modeled. Therefore, in order to quantify uncertainty about an unsampled value, it is important first and foremost to quantify the important spatial relationships, that is, we must quantify or model the underlying spatial continuity.

The simplest possible quantification consists in evaluating the correlation coefficient between any datum value measured at a location $u = (x, y, z)$ and any other measured a distance h away. Providing this correlation for various distances h will lead to the definition of a correlogram or variogram, which is one particular spatial continuity model discussed.

In the particular case of modeling the subsurface, spatial continuity is often determined by two major components: the structural features, such as fault and horizon surfaces, and the continuity of the properties being studied within these structures. Since these two components present themselves so differently, the modeling of spatial continuity for each is approached differently, although it should be understood that there may be techniques that apply to both structure and properties. Modeling of structures is treated in Sect. 6.

The models discussed in this section are stochastic models (as opposed to physical models) and are used to model so-called static properties such as a rock property or a soil type. They are rarely used to model dynamic properties (those that follow physical laws) such as pressure or temperature, unless it is required to interpolate pressure and temperature on a grid or do a simple filtering operation or statistical manipulation. Dynamic properties follow physical laws, and their role in modeling uncertainty is not discussed in this paper, but the reader is referred to [9] and references therein.

Three spatial models are presented in this section: (1) the correlogram/variogram model, (2) the object (Boolean) model, and (3) the 3D training image models. The variogram is a model based on mathematical considerations rather than physical ones. While the variogram may be the simplest model of the set, requiring only a few parameters, it may not be easy to interpret from sparse data, nor can it deliver the complexity of real spatially varying phenomena. Both the object-based and the training image-based models provide a more realistic perspective. They require a prior thorough understanding, or interpretation, of the spatial phenomena and many more parameters. Naturally, any interpretations are themselves subject to uncertainty.

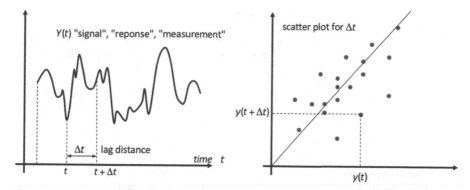

Fig. 2 A 1D time series (*left*). Calculating the correlation coefficient for a given lag distance or time interval $\triangle t$ (*right*)

3.1 The Variogram

3.1.1 Autocorrelation

To construct a variogram for an observable variable $Y(t)$, we first compute correlation coefficients for $Y(t)$ with a shifted version of itself $Y(t + \triangle t)$, which we denote by $\hat{Y}(t)$. We expect that $Y(t)$ and $\hat{Y}(t)$ are generally strongly associated for small $\triangle t$ and are weak for large values of $\triangle t$. An example scatter plot of observations $(Y(t), \hat{Y}(t))$, or $(Y_i, \hat{Y}_i)$, for some fixed $\triangle t$ and $N - 1$ observation pairs, is shown in Fig. 2. The correlation coefficient r_Y, which is a function of $\triangle t$, is now computed as

$$ r_Y(\triangle t) = \frac{1}{N-1} \sum_{i=1}^{N-1} \left(\frac{Y_i - \bar{Y}}{s_Y} \right) \left(\frac{\hat{Y}_i - \bar{\hat{Y}}}{s_{\hat{Y}}} \right), $$

where s_Y and $s_{\hat{Y}}$ are the standard deviations for the Y and $\hat{Y}$ series respectively.

Calculating the correlation coefficient for various intervals $\triangle t$ and plotting $r(\triangle t)$ versus $\triangle t$ results in a correlogram or autocorrelation plot. For $\triangle t = 0$, the correlation is naturally $r(0) = 1$, since $Y(t)$ is perfectly correlated with itself.

The same idea can be extended to multiple dimensions. For each direction in space, we can create correlograms following the above procedure, as illustrated in Fig. 3. We do this for a set of directions. The directions are important, because spatial phenomena, particularly in the earth sciences, are often oriented according to a preferential direction.

3.1.2 The Variogram and Covariance Function

While the empirical correlation function is a perfect way to characterize the spatial or temporal degree of correlation between events distributed in space or time, it

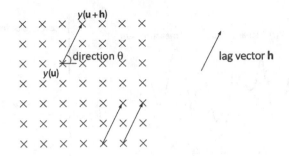

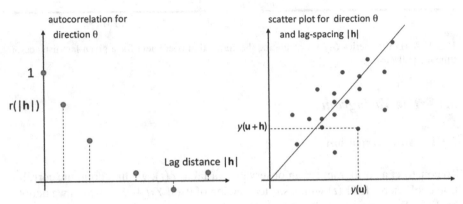

Fig. 3 Direction and lag spacing between two samples on a regular grid, corresponding scatter plot and construction of the correlogram or autocorrelation plot for that direction. Here u is a spatial coordinate

is not the traditional way to do so in the Earth and environmental sciences or in geostatistics [8, 14, 15]. In geostatistics, one prefers working with the variogram instead, the reason for which is explained below.

Assume that we are studying a variable Z that varies in space and/or time, that is, it is associated with a coordinate $u = (x, y, z)$ or $u = (x, y, z, t)$. We compute its autocorrelation function $\rho(h)$, and express it in terms of expected values E and variance $Var(Z)$. We have

$$\rho(h) = \frac{E\left[(Z(u) - m)(Z(u + h) - m)\right]}{Var(Z)},$$

with $m = E[Z(u + h)] = E[Z(u)]Var(Z)$. Clearly, $\rho(h) = 1$ for $h = 0$. Here, h again is a vector with a given lag spacing and direction, as shown above; $\rho(h)$ is computed for a finite set of vectors h. For other lags and directions, results are interpolated or extrapolated, which must be done carefully.

We now introduce a measure of correlation similar to the autocorrelation function, which is termed the covariance function $C(h)$ and is given by

$$C(h) = E\left[(Z(u) - m)(Z(u + h) - m)\right].$$

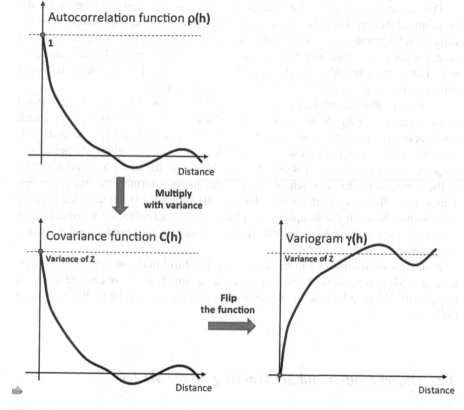

Fig. 4 Relationship between autocorrelation, covariance, and variogram functions

That is, $C(h)$ is simply $\rho(h)$ multiplied by $Var(Z)$ and hence contains information about the variance of Z.

We next define the variogram as

$$\gamma(h) = Var(Z) - C(h) = \frac{1}{2}E\left[(Z(u) - Z(u+h))^2\right].$$

A variogram can be seen as a measure of the *geological distance* between any two points in space as opposed to the Euclidian distance.

The relationships between $\rho(h)$, $C(h)$, and $\gamma(h)$ are illustrated in Fig. 4. Note that the equation for the variogram does not call for the calculation of a mean or variance, because it can simply be estimated from a limited sample as

$$\gamma(h) \approx \frac{1}{2n(h)}\sum_{u}(z(u) - z(u+h))^2,$$

with $n(h)$ the number of pairs found for the given lag h.

The advantage of using the variogram $\gamma(h)$ over the autocorrelation or the covariance function is that it is a more stable function. Suppose, for example, that the variance is large and grows as more data are gathered over an increasing area. In that case, the covariance function, which starts at the variance, would become unstable and change continuously. The variogram does not have this problem. It would simply start at zero for $h = 0$, and then will keep increasing.

In general, less correlation is expected as h increases. This would correspond to an increase in $\gamma(h)$. At some point, $\gamma(h)$ flattens out. The distance h at which this happens is termed the range, and the level of the plateau is referred to as the sill. The sill is often found to be close to the variance of the sample values. By definition, the variogram value at $h = 0$ is exactly zero. However, for small h, a sudden jump in the variogram value is often observed. This jump is termed the nugget effect. This nugget effect is often due to small-scale variability that is not sampled because the distance between the samples is too large. Another reason for a nugget effect is error (noise) in the measurements. If these errors act like random noise, then a nugget effect will be observed.

Although variograms are more stable and insightful than autocorrelations, they remain crude descriptions of actual spatial phenomena and cannot capture their full complexity. We next look at the object models that are designed with this particular objective.

3.2 Object Models and 3D Training Image Models

A variogram captures spatial continuity by considering only sample values taken two at a time. It is clear that it cannot capture complex spatial phenomena, such as the sinuous variation of a channel or the growth of carbonate mounds and reefs that may need hundreds of parameters for a complete description rather than the few available in variograms, including nuggets, sills, and ranges. Figure 5 depicts a couple of examples.

Object models (also termed Boolean models) were introduced to overcome some of the limitations of the variogram-based tools by importing realistic shapes and associations into a model. Crisp curvilinear shapes are often hard to model with cell-based techniques. The object simulation approach consists in dropping directly onto the grid a set of objects representing different categories (rock types, sedimentary facies, and fractures). These objects are then moved around to match the data by Markov chain simulation [4, 20]. This technique has mostly been used to model sedimentary objects in reservoirs or aquifers, but many other applications can be envisioned, such as, for example, the simulation of gold veins. The first step in object modeling is to establish the various types of objects, e.g., sinuous, elliptic, or cubic, and their dimensions, including width, thickness or width to thickness ratio, vertical cross section parameters, and sinuosity. The parameters can vary according

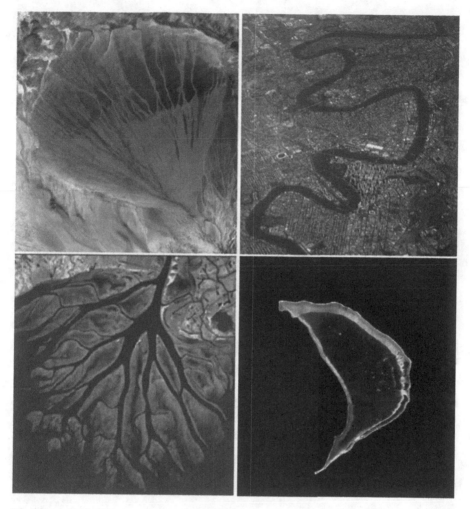

Fig. 5 An alluvial fan (*top left*), a meandering river (*top right*), a delta (*bottom left*), an atoll (*bottom right*)

to a user-specified distribution function. Next, the spatial relationships between objects are defined. This may include erosion of one object by another, embedding, or attraction/repulsion.

In many cases, it is not possible to capture spatial complexity by a few variogram parameters or even by a limited set of object shapes. The 3D training image approach is a relatively new tool for modelers to communicate the spatial continuity explicitly as a full 3D image. It is a conceptual rendering of the major variations that may exist in the studied area, akin to the texture mapping approach used in the games industry. An illustration is given in Fig. 6. A particular pattern is presented, in this case fluvial. The available data are then fit to the training image to create a

Training images

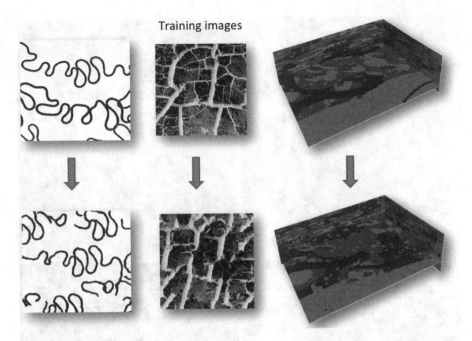

Fig. 6 A few example training images and Earth models produced from them exhibiting similar patterns

representative Earth model. Training images may be defined at various scales, from the large 10–100 km basin scale to the pore scale. Often, many alternative training images are created, reflecting uncertainty about the understanding of the studied phenomenon.

The design of object models and training images under uncertainty is the topic of the next section.

4 Modeling Spatial Uncertainty with Training Images

A variogram model, Boolean model, or 3D training image provides a model for the spatial continuity of the phenomenon being studied. In this section we discuss methods for generating Earth models that reflect what is captured in the chosen spatial continuity models. Typically, Earth models are not unique, and many models that represent the same spatial continuity can be created. The set of models thus generated represents a type of uncertainty that is termed spatial uncertainty or model of spatial uncertainty.

We limit ourselves in this overview paper to the training image approach. Variogram-based methods and Boolean modeling approaches are not discussed here. They are often used, however, and we refer the reader to [9] for further

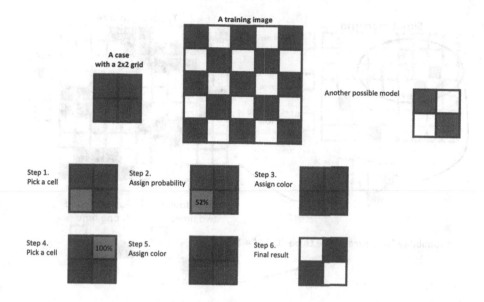

Fig. 7 An example of sequential simulation with training images

reading. In object-based algorithms, multiple Earth models are generated by fitting objects to the available data on the underlying Earth model grid. This is typically done in an iterative fashion to minimize the mismatch between the chosen object parameterization and the actual data. Iterative schemes are typically tailored to the object parameterizations and are often computationally intense. We refer to [4, 9, 20] for additional information.

4.1 Sequential Simulation with Training Images

Most of the geostatistical tools for modeling spatial uncertainty are cell-based (pixel-based). Among many of the methods available, sequential simulation is attractive, because it scales to the millions of grid points often used in Earth models. The idea behind sequential simulation is straightforward: starting from an empty Cartesian grid, a model is built one cell at a time by visiting each grid cell along a random path, assigning values to the grid cell, and continuing until all cells are visited. Regardless of how grid cell values are assigned, the value assigned to a grid cell depends on the values assigned to all the previously visited cells along the random path and the chosen training images. It is this sequential dependence that forces a specific pattern of spatial continuity into the model.

Consider the simple example of Fig. 7. Here, the goal is to generate a binary model for a 2×2 grid that displays a checkerboard pattern similar to the training

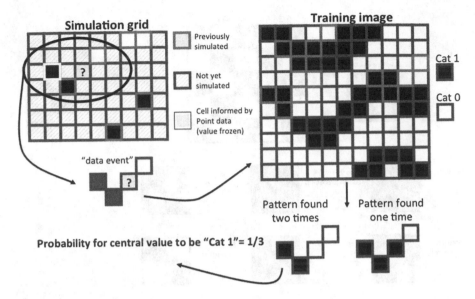

Fig. 8 Second example of sequential simulation with training images

image shown at the top of the figure. Note that this training image has 13/25 dark cells and 12/25 light cells. We pick any of the four cells. Using a $13/25 = 52\%$ probability for a dark cell and a random drawing, we determine whether to draw the cell dark or white. Suppose here that the outcome is dark. We then pick the next cell randomly. We determine the color of this cell using the training image. Clearly, the training image pattern dictates that this cell should be dark also and that the remaining cells are then to be colored white.

By changing the order in which cells are visited, or by changing the random drawing, different results will be obtained. For this particular case, there can be only two possible final Earth models, each with an almost equal chance of being generated. In geostatistical jargon, we also refer to these Earth models as realizations.

Another example is given in Fig. 8. The probability of the central cell being channel sand, given its specific set of neighboring sand and no-sand data values, is calculated by scanning the training image for occurrences of this data event: three such events are found, of which one yields a central sand value; hence the probability of sand is $1/3$. By random drawing, a category is assigned. This operation is repeated until the grid is full. This procedure results in a simulated Earth model, or realization, which will display a pattern of spatial continuity similar to that depicted in the training image.

In actual cases, data provide local constraints on the presence of certain values/categories. In sequential simulation, such constraints are handled easily by assigning category values to all grid cells that have given point data, that is, data that directly inform the grid cell value. The cells containing such constraints are never

Facies type	Conceptual description	Stratigraphy	Length (m)	Width (m)	Thickness (ft)
Tidal bars	Elongated ellipses w/ upper sigmoidal cross-section	Anywhere	2000-4000	500	3-7
Tidal Sand flats	Sheets (rectangular)	Anywhere Eroded by sand bars	2000	1000	6
Estuarine sands	Sheets (rectangular)	Top of the reservoir	4000	2000	8
Transgressive lags	Sheets (rectangular)	Top of the estuarine sands	3000	1000	4

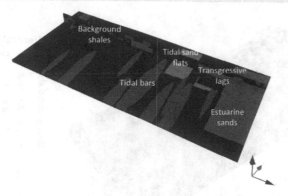

Fig. 9 Table with rock type relations and geometric description. The corresponding training image was generated using an unconstrained object simulation method

visited and their values never reconsidered. The sequential nature of the algorithm forces all neighboring simulated cell values to be consistent with the data.

A more elaborate example is that depicted in Fig. 9, in which rock types in a tidal dominated system are modeled. The Earth model is a $149 \times 119 \times 15$ grid, with each grid of size $40m \times 40m \times 1m$. The model includes five rock types: shale (50%), tidal sand bars (36%), tidal sand flats (1%), estuarine sands (10%) and transgressive lags (3%). Using an unconditional object-based method a training image was constructed using the rules that tidal sand flats are eroded by sand bars and transgressive sands always appear on top of estuarine sands. Additionally, trend information is available from well-logging and seismic data. Trend information is usually not incorporated in the training image itself but instead used as additional constraints for generating an Earth model: the training image reflects only the fundamental rules of deposition and erosion. Trend information available here gives that sand bars and flats are expected to prevail in the southeast part of the domain, and that shale is dominant at the bottom of the model, followed by sand bars and flats, whereas estuarine sands and transgressive lags prevail in the top part. Using the point data obtained from 140 wells, an aerial proportion map and a vertical proportion curve are estimated for

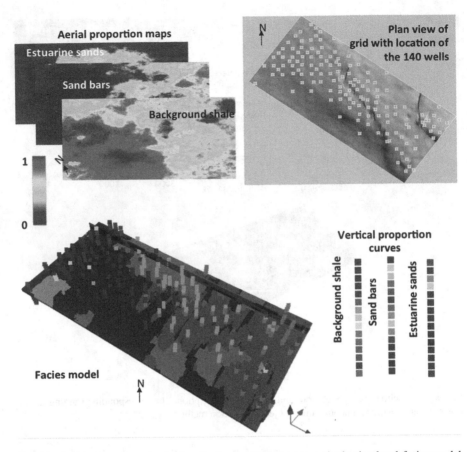

Fig. 10 Aerial proportion map and vertical proportion curves, a single simulated facies model constrained to data from 140 wells and reflecting the structure of the training image shown in the previous figure

each category. Figure 10 shows a single Earth model generated using this approach. The model of the imposed erosion rules, depicted by the training image, matches exactly all data from 140 wells and follows the trends described by the proportion map and curves.

5   Constraining Spatial Models of Uncertainty

Many different sources of information are typically available for modeling a property or variable of interest in the earth sciences. The question then is how to combine these sources of information to model the spatial variable of interest. Ideally, the more data we have, the smaller the uncertainty in the variable of interest, but this depends on how much information each data source carries about the unknown and how redundant this source of information is with respect to other data sources in determining this unknown. Thus far, we have considered direct measurements of the variable of interest at the scale required as well as spatial continuity information. Other data sources include remote sensing data or geophysical measurements, among others. A common problem in building Earth models and constraining models of uncertainty lies in combining data sources that are indirect and/or at a different scale from the modeling scale with data that provide more direct information, such as those obtained through sampling. We discuss the probabilistic approach in this section. For inverse modeling approaches, we refer to [23].

5.1 Probability-Based Approach

We classify the available data in two groups: samples and geophysical surveys. By samples we mean a detailed analysis at a small scale. These could, for example, be soil samples, core samples, plugs, well-logging, point measurements of pressure, or air pollution. By geophysical measurements we understand the various remote sensing techniques applied to the Earth to gather an image of the Earth or surface being modeled, such as synthetic aperture radar, seismic, or ground-penetrating radar. There may be various geophysical data sources (electromagnetic, seismic, gravity) and various point sources. Other data available may be soft data that give qualitative information about the Earth model. The main challenge is to use the data so that uncertainty of the Earth model is reduced. Added complexity comes from incomplete or partial data and the need to combine data across various scales.

We proceed in two steps in order to include these data in our model of uncertainty. The first step is a calibration step, in which we ask how much information is contained in each data source. The second step is the integration step, in which we ask how to combine these various sources of information content into a single model of uncertainty.

The amount of information contained in a data source is dependent on many factors, such as the measurement configuration, the measurement error, the physics of the measurement, and the scale of modeling. To ascertain the information content of a data source, we can use the conditional distribution discussed above. The conditional probability $P(A|B)$ models the uncertainty of some target variable A, given some information B. In our case, B is the data source and A is the model parameter. To determine this conditional probability, which we refer to as the calibration function, we need data pairs (a_i, b_i), that is, at some limited set of locations it is necessary to have observed the true Earth as well as the data source. This is possible in many applications. For example, in reservoirs, we can obtain pairs of porosity and seismic impedance measurements at the well locations.

Consider now the situation in which several such calibrations have been obtained because many data sources are available. In other words, several $P(A|B_1), P(A|B_2), \ldots$ have been found. The next question is how to combine these to find $P(A|B_1, B_2, \ldots)$. One way is to perform a single calibration involving all the data sources at once to get this combined conditional probability, but this is often too difficult or it would require a high-quality and rich-calibration data set. Another way is to use input from experts from different fields. Here, we provide a simple and quite general way of combining these probabilities.

Consider first a very simple binary problem. Suppose we have two sources of information (events B_1 and B_2) that inform us that the chance it will rain tomorrow (event A) is significant. From the first source B_1 we deduce that there is a probability of 0.7 of event A occurring. From the second source B_2 we obtain a probability of 0.6. We also know that the historical probability of it raining on the day that is tomorrow is 0.25. We note that this prior or background conditional probability is lower than the probabilities given by the two sources. When we now ask about the probability of it raining tomorrow, we have also to think about redundancy. This is difficult to measure and requires making assumptions about the nature of the information sources with respect to the target unknown. We can simply start by making an assumption or hypothesis and then check whether the results we get are reasonable. For example, we can assume that the relative contribution of information B_2 to knowledge of A is the same regardless of the fact that we have information B_1.

We now define

$$b_1 = \frac{1 - P(A|B_1)}{P(A|B_1)}, \quad b_2 = \frac{1 - P(A|B_2)}{P(A|B_2)}, \quad a = \frac{1 - P(A)}{P(A)}.$$

Each of these scalar values can be seen as a distance. For example, if $b_1 = 0$, then $P(A|B_1) = 1$, and hence B_1 tells all about A, while if $b_1 = \infty$, then B_1 ensures that A will not happen. What we would like to know is

$$x = \frac{1 - P(A|B_1, B_2)}{P(A|B_1, B_2)},$$

which gives

$$P(A|B_1, B_2) = \frac{1}{1 + x}.$$

Since

$$\frac{b_2 - a}{a} = \frac{x - b_1}{b_1},$$

we find that

$$x = \frac{b_1 b_2}{a}, \quad \text{and} \quad P(A|B_1, B_2) = \frac{a}{a + b_1 b_2}.$$

In our example, this gives $P(A|B_1, B_2) = 0.91$, which suggests that we have more certainty about it raining from these two data sources than from considering each data source separately. In other words, the model assumption enforces compounding, or reenforcement, of information from the two sources.

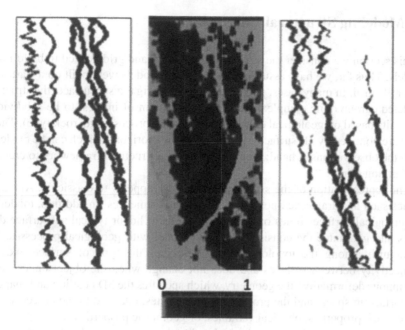

Fig. 11 Training image (*left*), probability derived from seismic (*middle*), Earth model (*right*)

Next, we consider how to use this model of data redundancy in the context of spatial uncertainty. Consider the example of geophysical data and consider determining the presence or absence of sand at each grid block in an Earth model. Consider that we have determined the function $P(A = \text{sand}|B_2)$, with B_2 the geophysical information. This means that at every location in the grid where we have a geophysical measurement, this measurement can be replaced with a conditional probability by means of the calibration function. Figure 11 gives an illustration. The color in the middle plot represents the probability of sand for the given geophysical information (in this case, seismic data, which are not shown). The spatial continuity is expressed in the training image on the left. We now create an Earth model that contains more sand channels in areas where the probability of sand is higher. This is simply done by means of an extension of the sequential simulation methodology. In the algorithm described in the previous section, we first assign any hard data to appropriate cells in the simulation grid. Then, we define a random path to loop over all grid cells. At each grid cell, we now determine the uncertainty of that unsampled grid value given the data an the previously simulated cell values in terms of a probability distribution. We do this in three steps: (1) determine $P(A|B_1)$ from the training image by scanning; (2) determine the probability $P(A|B_2)$ from the sand probability at the current grid location; (3) use the equations given above to determine $P(A|B_1, B_2)$, with $P(A)$ the proportion of sand. We now sample a value from the obtained probability distribution by Monte Carlo simulation and assign it to the grid cell. The right-hand figure in the illustration shows an example Earth model created with this straightforward extension of sequential simulation.

6   Modeling Structural Uncertainty

In this section we consider the structural framework and grid needed to build Earth models. Thus far, we have assumed that the Earth models were built on a Cartesian 2D or 3D grid. In many cases, particularly in modeling the subsurface, the structure modeled can be quite complex, requiring the domain of interest to be subdivided into units based on geological considerations (setup process, age, rock type). These units are bounded by 3D surfaces such as faults and horizons, which can be modeled from the observations at hand; they then serve as a structural framework to create a conforming grid.

The representation of the structural framework appeals to classical CAD techniques such as parametric lines and surfaces and/or meshes. In addition, modeling of geological surfaces relies on specific methods to honor typical subsurface data types and guarantee the consistency of the model with geological processes. We distinguish among the topology, which describes the type of surface and the connectivity between surfaces and does not change when the object undergoes a continuous deformation; the geometry, which specifies the 3D position and shape of the surface in space; and the properties, or attributes, attached to the object, which can be rock properties, physical variables, or geometric properties.

Here, we focus on the uncertainty related to the geometry and to the topology of structural models as properties were handled in earlier sections. Data most used for structural modeling are geophysical images obtained through geophysical surveys such as seismic or electromagnetic surveys. Geophysical data provide complete coverage of the subsurface along one section (e.g., 2D seismic) or over a whole volume (e.g., 3D seismic). The geophysical data used are the outcome of a complex chain of geophysical processing. For example, seismic acquisition is based on the emission of an artificial vibration (the source) either onshore or offshore, whose echoes are recorded by a set of geophones. The seismic waves emitted by the source undergo refraction and reflection when propagating through rocks of different nature. This signal carries the contrasts of impedance (the product of seismic wave velocity by the rock density) as a function of time (the travel time between the source and the geophone). Seismic processing turns the raw data into a usable 3D seismic image. The processing of seismic data is very complex and computationally very demanding, requiring a number of corrections and filtering operations, whose parameters are generally inferred from the raw seismic data and calibrated at wells. For example, the conversion of travel times to depth is ambiguous, since it requires an estimation of seismic velocities that are not known a priori. Therefore, seismic data are not precise, especially in the vertical direction. Also, they have a poor resolution (from a few meters for shallow, high-resolution seismic to about 20–50 m for classical surveys). Lastly, the significance of seismic images decreases as depth increases, due to the attenuation of the signal amplitude. Notwithstanding these limitations, the value of seismic data is its ability to cover an entire volume. The fuzzy picture provided by seismic data is of paramount importance for structural modeling, since it provides a view of 3D geometries. Seismic amplitudes are

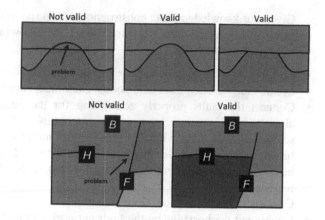

Fig. 12 Valid and invalid configurations of faults and horizons in the subsurface

routinely used to extract significant surfaces such as horizons, unconformities, and faults. Since the seismic image is fuzzy, such extraction requires tedious and usually manual interpretations.

6.1 Creating a Structural Model

Discontinuities within the subsurface are due to changes of depositional conditions, erosion, tectonic events such as faulting and folding, or late transport of reactive fluids. In designing a structural model, surfaces of discontinuity cannot be modeled one at a time, but must be modeled simultaneously, since the relation between the surfaces must be taken into account. The modeling is further complicated by often sparse and also noisy data, which necessitate consistency checks to ensure that the structural models generated are geologically valid. An example of a consistency check is that a surface must not self-intersect. Also, volumes defined by a set of interfaces must not overlap. Then, the succession of geological processes imposes a hierarchy on the geological objects. One consequence is that an interface between two rock volumes is necessarily bounded by other interfaces. For instance, a surface representing a horizon cannot float in space: its borders have to lie on other natural surfaces, such as a fault or domain boundary surface. The only exception to that rule is faults: the borders of a fault surface can float in space. Some example of valid and invalid surfaces are illustrated in Fig. 12.

The criteria defined above are only necessary conditions for enforcing geological validity. Defining sufficient conditions is much more difficult and calls for higher-level concepts in structural geology and sedimentology. Enforcing such conditions intrinsically in the model-building process is not always easy or applicable, and it requires a careful prior geological analysis of the domain being studied. Alternatively, models can be checked for consistency with some type of data or physical principles. This a posteriori check is usually computationally expensive.

Given our knowledge about construction of a surface and the various conditions required to build a consistent structural model, the following basic steps can be used to build such a model:

- Collect a data set on fault and horizon surface, e.g., from seismic data.
- Create fault surfaces from data points, each independently at first.
- Connect the faults properly accounting for the geological consistency rules above. Significant uncertainty is present at this stage, because subsurface data seldom provide a clear image close to discontinuities. It is up to the modeler, based on analogous reasoning and careful data analysis (e.g., regional tectonic history, hydraulic connectivity between wells), to decide on the fault connection patterns.
- Construct horizon surfaces from data points.
- Cookie-cut the horizons by the fault network and update the horizon geometry inside each fault block.
- Merge fault and horizon surfaces.

The logic behind these steps follows the genetic rules of most subsurface structures, that is, horizons (layers) were created first, and only then were these layers deformed and faults created. Hence faults are assumed younger in time and should be constructed first.

6.2   Modeling Structural Uncertainty

Knowing how we can build structural models brings us to the question how we can build multiple structural models that serve as a representation or model of uncertainty. This is not as trivial a question as the modeling of properties on a regular grid. The various constraints of geological consistency as well as the difficulty of automating the construction of structural models make this a difficult task.

The main sources of uncertainty lie in the data source (seismic) being used for structural modeling and the interpretation of such data by a modeler. The order of magnitude of uncertainty can be different depending on the data acquisition condition in the field (i.e., land data or marine data, 2D or 3D seismic), the subsurface heterogeneity, and the complexity of the structural geometry. For example, land data generally provide poorer seismic data than marine data.

Although it is difficult to set a universal rule, a typical example of hierarchy in subsurface structural uncertainty based on seismic data is suggested in the overview of Fig. 13.

Seismic data basically measure a change of amplitude of a seismic wave in time at a certain position. This signal is indicative of the contrast of rock impedance at a certain depth in the subsurface. A first source of uncertainty in structural interpretation occurs when two different rock units have a small impedance contrast (e.g., granite and gneiss). Then, to go from time to depth, it is necessary to know the seismic velocity (velocity × time = depth), that is, how fast these waves travel

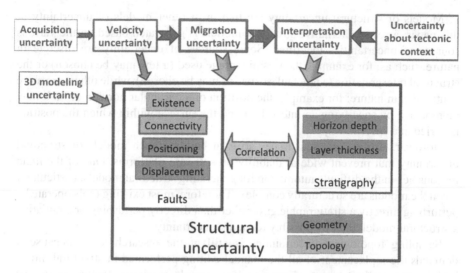

Fig. 13 Main factors and types of structural uncertainty, with typical significance figured as frame and arrow width

through the subsurface. This velocity needs to be somehow determined, since it is not measured directly, and hence this is subject to uncertainty. The entire process of putting all the recorded seismic signals in their correct place is called migration. Structural uncertainty resulting from uncertain migration can be a first-order structural uncertainty, especially when seismic data are of poor quality. In such situations, multiple seismic images migrated using different velocity models can produce significantly different structural interpretations that exhibit different fault patterns and appearance or disappearance of faults depending on which seismic image is considered.

With poor seismic data, a single seismic image can produce considerably different structural interpretations depending on the different decisions made on horizon/fault identification. Such an uncertainty from structural interpretation can be of first order, especially when the structural geometry is complex, since multiple interpretations can exhibit a significant difference in number of faults or fault pattern because of human bias in interpretation. This uncertainty is modeled by providing multiple possible alternatives of structural interpretations. These may be defined within one or several tectonic scenarios.

Correlating horizons across a fault can be difficult unless wells are available on both sides of the fault. Erroneous horizon identification would lead to misinterpreting the fault throw, which is the vertical component of the fault slip. The magnitude of uncertainty related to fault positioning depends on the quality of the seismic image on which the interpretation is done. Its magnitude is evaluated through visual inspection of the seismic image. This uncertainty is also of lower-order importance compared to the uncertainty related to the fault identification or the uncertainty in fault connectivity. It usually increases with depth, and may be locally smaller if the fault has been identified along boreholes.

Models of structural uncertainty are like most other models of uncertainty in this paper: a set of alternative structural models is generated based on the various sources of uncertainties identified. Some sources of uncertainty are discrete in nature, such as, for example, the seismic image used (a few may be chosen) or the structural interpretation (a few fault scenarios may be chosen), while others are more continuous in nature: for example, the position of a particular horizon or fault may be modeled by specifying an interval around the surface within which the position is varied in a specific way.

However, various practical issues arise in building such models of structural uncertainty that prevent wide availability on software platforms. One of the main reasons lies in the difficult automation for generating structural models, particularly when the models are structurally complex. Therefore, most existing tools operate by perturbing directly a stratigraphic grid rather than directly perturbing or simulating a structural model, thereby possibly reducing uncertainty.

Sampling topological uncertainties is still at the research level, because it demands the replacement of all the manual editing performed in structural model building by ancillary expert information. For example, stochastic fault networks can be generated from statistical information about fault orientation and shape, relationships between fault size and fault displacement, and truncation rules between fault families.

References

1. CAERS, J. AND ZHANG, T., *Multiple-point Geostatistics: a Quantitative Vehicle for Integrating Geologic Analogs into Multiple Reservoir Models*, in Integration of outcrop and modern analog data in reservoir models (eds G.M. Grammer, P.M. Harris and G.P. Eberli), AAPG Memoir 80, American Association of Petroleum Geologists, Tulsa, OK, 383–394, 2004.
2. DEUTSCH, C.V. AND JOURNEL, A.G., *GSLIB: The Geostatistical Software Library*, Oxford University Press, 1998.
3. HALDERSON, H.H. AND DAMSLETH, E., *Stochastic Modeling*, Journal of Petroleum Technology, 42(4), 404–412, 1990.
4. HOLDEN, L., HAUGE, R., SKARE, Ø., AND SKORSTAD, A., *Modeling of Fluvial Reservoirs with Object Models*, Mathematical Geology, 30, 473–496, 1998.
5. ISAAKS, E.H. AND SRIVASTAVA, R.M., *An Introduction to Applied Geostatistics*, Oxford University Press, 1989.
6. RIPLEY, B.D., *Spatial Statistics*, John Wiley & Sons, Inc., New York, 2004.
7. BARDOSSY, G. AND FODOR, J., *Evaluation of Uncertainties and Risks in Geology*, Springer, 2004.
8. CAERS, J. , *Petroleum Geostatistics*, Society of Petroleum Engineers, Austin, TX, 2005.
9. CAERS, J., *Modeling Uncertainty in the Earth Sciences*, John Wiley & Sons, Inc., 2011.
10. BORRADAILE, G.J., *Statistics of Earth Science Data*, Springer, 2003
11. DAVIS, J.C., *Statistics and Data Analysis in Geology*, John Wiley & Sons, Inc, 2002.
12. ROHATGI, V.K. AND EHSANES SALEH, A.K. MD. , *An Introduction to Probability and Statistics*, John Wiley & Sons, Inc., 2000.
13. CHILES, J.P. AND DELFINER, P., *Geostatistics: Modeling Spatial Uncertainty*, John Wiley & Sons, Inc, 1999.

14. DALY, C. AND CAERS, J., *Multiple-point Geostatistics: an Introductory Overview*, First Break, 28, 39–47, 2010.
15. HU, L.Y. AND CHUGUNOVA, T., *Multiple-point Geostatistics for Modeling Subsurface Heterogeneity: A Comprehensive Review*, Water Resources Research, 44, W11413. doi:10.1029/2008WR006993, 2008.
16. LANTUEJOUL, C., *Geostatistical Simulation*, Springer, 2002.
17. BOND, C.E., GIBBS, A.D., SHIPTON, Z.K., AND JONES, S., *What do you think this is? "Conceptual Uncertainty" in geoscience interpretation*, GSA Today, 17, 4–10, 2004.
18. CAUMON, G. , *Towards Stochastic Time-varying Geological Modeling*, Mathematical Geosciences, 42(5), 555–569, 2010.
19. CHERPEAU, N., CAUMON, G., AND LEVY, B., *Stochastic Simulation of Fault Networks from 2D Seismic Lines*, SEG Expanded Abstracts, 29, 2366–2370, 2010.
20. HOLDEN, L., MOSTAD, P., NIELSEN, B.F., *Stochastic Structural Modeling*, Mathematical Geology, 35(8), 899–914, 2003.
21. SUZUKI, S., CAUMON, G., AND CAERS, J., *Dynamic Data Integration into Structural Modeling: Model Screening Approach Using a Distance-Based Model Parameterization*, Computational Geosciences, 12, 105–119, 2008.
22. THORE, P., SHTUKA, A., LECOUR, M., *Structural Uncertainties: Determination, Management, and Applications*, Geophysics, 67, 840–852, 2002.
23. TARANTOLA, A., *Inverse Problem Theory, and Methods for Model Parameter Estimation.* Society for Industrial and Applied Mathematics, 2005.

Printed in the United States
By Bookmasters